The DNA Detectives

The DNA Detectives

W.J. WALL

ROBERT HALE · LONDON

© W.J. Wall 2005
First published in Great Britain 2005

ISBN 0 7090 7504 9

Robert Hale Limited
Clerkenwell House
Clerkenwell Green
London EC1R 0HT

The right of W.J. Wall to be identified as
author of this work has been asserted by him
in accordance with the Copyright, Designs and
Patents Act 1988

A catalogue record for this book is available
from the British Library

2 4 6 8 10 9 7 5 3 1

Typeset in 11/14.5pt Century Schoolbook
Printed by St Edmundsbury Press Limited
and bound by
Woolnough Bookbinding Limited

CONTENTS

1 A BRIEF HISTORY OF PERSONAL IDENTIFICATION

It has always been necessary since human beings were sentient for one human being to be able to identify another. This extends, quite naturally, into the animal kingdom. With human beings, it makes no sense for an animal that mates by choice to be unable to recognize their mate; the result of being unable to recognize an individual would be random mating. So, given the need to identify an individual, we might naturally ask, how is it done? The method of identifying a person most people would say that they use is based not upon a technique, or feature, but on a composite process. Included in the process are all the five senses. If you cannot ask the individual themselves who they are, as in the case of a dead body, what you do is ask someone who knows them, assuming, of course, that you trust the person you ask to make the identification. For centuries this was the only way identification could be made – there were no sophisticated methods to call on. For some there might have been an item of jewellery, say, a signet ring that would have allowed identification, but nothing that could be relied upon absolutely.

The apparently simple action of identifying an individual does not give a clear idea of just how complicated the process really is. It is all carried out in a split second and is the result of a staggeringly intricate set of actions within our brains. It is possible for two close companions to identify each other over a distance in which it is impossible to see specific features. But this is done by looking at movement, scale and shape in a composite manner. This is all the more amazing since our eyes are not, optically, the best in the animal kingdom; in fact they are rather poor, although the spatial resolution, the visual acuity, is really quite surprising. However, what they lack in optical clarity, humans more than make up for in neural processing.

To identify a person by sight is difficult to explain, but not difficult to do. What we experience here is the gap between the ease of stating something that we all do, and the difficulty of explaining the science behind it. This is one of the things that this book will attempt to reconcile. Trying to put down in a systematic and rational manner all the parts of the process of saying 'yes, that's him, I recognize him' would involve pages and pages of lists. Yet we all do it in a heartbeat.

What we do when we recognize someone by sight is a complex activity based on all manner of things from the way they walk to the way they stand and hold their posture. So how do we address the problem of personal recognition? There are two things to be said about this at the outset. The first is that there is still no substitute for being able to ask a third party the question 'Is this the person I think he or she is?' Interestingly this has simple, but direct, consequences for security, both state and commercial. The best way of restricting access to any facility is simply to have a security officer on reception. If you are not recognized by the security officer, you do not get in. As technologies advance, the incredibly complex act of personal recognition – that requires a *human* security officer – is replaced in the field of securities by such things as fingerprint ID or voice recognition. All such techniques have one thing in common – they rely on there being an original meeting or fingerprint with which comparison, either by the security officer or fingerprint machine, can be made.

Personal recognition becomes a different kind of problem when you introduce criminal activity. The theoretical difficulties in recognition suddenly become very real problems when there is a very important question – 'whodunit?' In cases where there is no eyewitness, nobody who can say 'yes that was the person', it is even more difficult. This is the normal situation in murder and in rape cases, for example. Eyewitness testimony by a traumatized victim may not hold up against seasoned barristers. In fact, eyewitness descriptions of individuals are notoriously unreliable: two people describing the same individual are quite likely to come up with two different descriptions of that person.

So in the context of a criminal situation how exactly can we track down the guilty? These are the kinds of criminal cases that are of primary importance in this book, followed closely by questions of paternity and inheritance. It could be said, in everyday life, paternity suits have a greater financial impact than any other field where DNA profiling is used, since they impinge on both inheritance and issues of maintenance, the latter of which may result in a father having to pay upkeep for nearly two decades. Paternity disputes are also of particular importance because they can cause such bad feeling within and between families.

Given that the most common method of identifying an individual is (and was) by sight, in the nineteenth-century near past it seemed logical to try to quantify the process of facial recognition. Nowadays such an idea may not seem fanciful because of our enormous computing power, but in fact back then it was not thought fanciful either.

Perhaps the first attempt to quantify appearance was not directly associated with identification so much as quantifying things such as a tendency towards debauchery. This was *phrenology* – reading the bumps and ridges of the head. It was started by Franz Joseph Gall (1758-1828) and carried on by his disciple Johan Spurzheim. In 1791 he published a treatise, in German, on using the shape of the skull to discover emotional characteristics. Phrenology became ever more popular until 1802 when the Tory government of the day, led by Henry Addington, decided that the practice had to stop because, as they put it, it was dangerous to religion.

After the débâcle of phrenology came Alphonse Bertillon. He was born in France in 1853, the son and brother of statisticians, and lived through turbulent times until his death in 1914. His father was Louis-Adolphe, the head of the Paris Bureau of Vital Statistics, or as we would call it 'Social Statistics'. Alphonse Bertillon became the chief of the identification bureau in Paris in 1880, where he developed what came to be known as Bertillonage.

His new technique of Bertillonage, or probably more accurately 'anthropometry', was introduced by him in 1882 as a means of personal identification. To say that it was a bizarre method of personal identification is not putting it too strongly, simply because it involved measurements which could so easily alter. It required accurate measurement of various parts of the body, with additional notes on any scars or marks that the individual carried. It was assumed that these measurements would be able to correctly identify an individual, but there was a flaw, besides the inherent inaccuracies and variations in the way the data was collected. The only way that Bertillonage could identify an individual was by already having the detailed measurements on file and comparing them with an arrested individual. This really serves no other purpose than to confirm that an arrested person is either who they say they are or who they said they were. Considering the types of measurements that are made using this system, it is surprising that any matches were ever found. If you take the following list of measurements that were part of the process, it helps to realize the flaws inherent in it.

Height standing
Height sitting
Reach from fingertip to fingertip
Length and width of head
Length and width of right ear
Length of left foot
Length of middle finger
Length of left little finger
Length of left forearm

The obvious observation that follows is that not only do we grow throughout our early and adolescent years, but that structures like ears and noses tend to keep growing throughout life, due to the cartilage within them. Even totally boned parts of the skeletal system that are measured in Bertillonage change as we grow older, such as the 'height standing' and 'height sitting'. The whole system was both contrived and unreliable, not least because taking the measurements is, without doubt, highly subjective. So when individuals were shown how to create a Bertillonage image of a person, they had to be thoroughly trained, but even so, measurements taken by one technician could not be directly compared with measurements taken by a different technician with any reliability or accuracy. The whole system was a mess, made up of a series of unrepeatable results. It was going to take a lot to make anything significant out of Bertillonage and in fact it was realistically never going to be much more than a curiosity of its time.

At the same time as Bertillon was developing his version of anthropometry, he was also developing his interest in handwriting and the methods by which it can be identified as originating from a specific person. It was this interest in handwriting that resulted in him being asked to give an opinion on who the writer of a certain document was. It was in 1894 that he got involved with the infamous Dreyfus case, which will serve as a good case study to begin with. It shows how the identification of individuals in criminal case's can be sadly wrong.

The background to the Dreyfus case was complicated and the outcome theatrical. It was so intricate a story with such an astonishing outcome that it would be hard to believe as fiction, let alone fact. Alfred Dreyfus was a French army officer who was born on 9 October 1859 in Mülhausen, Alsace, and was the son of a rich manufacturer. After studying at the *école polytechnique* he entered the army at the rank of lieutenant. Having successfully passed through the *école superieure de la guerre* (the staff college), he attained the rank of captain, receiving a position on the General Staff at the Ministry of War. However, during the summer of 1894 a letter was removed from the German Embassy that had been addressed to the German military attaché, Colonel Von Schwarzkoppen, from a

French individual. This letter did not itself contain information, but detailed the documents that would arrive at the Embassy for the attaché. So a French officer, or civil servant, was apparently betraying their country.

The writing on the letter bore a close resemblance to the handwriting of Alfred Dreyfus and so he was arrested on 15 October 1894. There was very little evidence, but the Minister of War, General Mercier, presented documents of various sorts to the court. These were later shown to be worthless as evidence. The presentation of this material to the court martial was carried out without the knowledge of either Dreyfus or his counsel. It was the identification of the handwriting as belonging to Dreyfus that clinched the conviction. This identification was given by Alphonse Bertillon. The resultant unanimous verdict, given on 22 December 1894, left Dreyfus with a life sentence exiled on Devil's Island, off the coast of Guyana. His sentence started on 15 March 1895. The family strived to get the case reviewed, the activity being led by his brother Mathieu, but this was impossible until new evidence could be produced.

It was the chance discovery by a third party that altered the situation with regards to Dreyfus. In 1896 Colonel Picquart, the head of information at the Ministry of War in Paris, came into possession of an apparently unsent letter from Colonel von Schwarzkoppen, the individual to whom the original, incriminating, letter had been sent. This letter had been found by a French agent, torn up in a waste paper basket. It was addressed to a French officer, Major Esterhazy, proving that he was in the pay of Schwarzkoppen. It turned out that Esterhazy was badly in debt, but there was more: the note that had been used to condemn Dreyfus was actually in Esterhazy's handwriting.

Here the story takes yet another twist. When Picquart presented his information to his superior officers, he was forbidden to pursue his enquiries, after which he was transferred to an administrative post in Tunisia. Before leaving for North Africa, however, he told his entire story to a lawyer friend, Leblois. It was Leblois who then discussed the case with the vice-president of the senate, Scheurer-Kestner. His political power and his conviction that Dreyfus was innocent finally started to re-open the case. At this same time, Mathieu Dreyfus, brother of Alfred, realized that Esterhazy had actually written the incriminating letter. When Mathieu wrote to the general staff on 15 November 1897 accusing Esterhazy, they were unwilling to acknowledge the problem, so Esterhazy was court marshalled and acquitted.

A most extraordinary series of events now took place with the press vehemently against the idea of a change in the sentence against Dreyfus. Indeed, it became so bad that Colonel Picquart was imprisoned for communicating confidential papers to a civilian, Leblois. But, in other quarters other opinions were being voiced. This was mainly amongst the French intelligentsia, not least with one individual, Emile Zola, now more famous for his works of fiction such as *Germinal* and *La Bête Humaine*. What he wrote in an open letter to the President of the Republic, written two days after the acquittal of Esterhazy, became famous: the renowned beginning of this letter was *'J'accuse'*.

Such was the power of this letter that he stirred up an entire government. At the instigation of the Ministry of War, Zola was tried and sentenced to imprisonment for a year, but he took refuge in England, rather than the enforced retirement at the expense of the French government. Eventually the Dreyfus case was laid before the court of appeal. This resulted in a new court martial being held, for which Dreyfus was brought back from Guyana. The trial lasted a month and to the surprise of an ever more interested public, resulted in a verdict generally thought odd. He was found guilty with extenuating circumstances and sentenced to ten years' imprisonment.

This story does get better for Dreyfus. Ten days later the French government pardoned Dreyfus. By pursuing the subject, sufficient evidence was available in 1903 to reopen the case, which resulted in the sentence being quashed in 1906. He was not only reinstated after this but during the First World War he attained the rank of Lieutenant Colonel and gained the Legion of Honour, after which he retired, until his death in 1935.

Such are the potential problems of trusting a single expert in a dubious area of pseudo-science: that is, a science that professes to give physical results without being backed up by empirical testing and data. Interpretation of handwriting is at best inexact, and, as we have seen, sometimes positively misleading. We trust the practitioner more than we should through a simple lack of understanding of the underlying basis of what is being presented in court.

There is another area where identification is important. What happens, for example, in a trade transaction, where one person wants to send an individual to a third party with a promise of money for goods, which were yet to be delivered? They sent a promissory note – however, it is quite evident that anyone could produce an IOU, so how could the recipient of the note be sure where it came from and that it was genuine? There had to be a simple method of certification of individuality. One way in which this problem could be addressed is by use of the fingerprint.

Although we now tend to think that fingerprint technology originated in the West, this is not so. In China, instead of sending money in payment for goods, a promissory note was sent, which could be exchanged for money. This was carried out in a very simple manner. If you imagine putting a thumbprint across the dividing line between a chequebook stub and cheque, when an individual presents the cheque for payment to you, if the thumbprint does not match then payment is refused. This is, in many ways, the same as having a series of numbers and signatures on a modern cheque, which is consequently paid by a third party.

In the case of Chinese notes it was a thumbprint that was used. This did not purport to identify an individual. It was simply the matching of two halves of a fingerprint, which was regarded as impossible to forge, which guaranteed the transactions. This recognition of the unique nature of fingerprints prompted the English engraver Thomas Bewick (1753–1828) to produce engravings of two of his fingertips, which he then used to authenticate his work.

Fingerprints have a definite aura of mystery about them. Even genetically identical individuals, such as identical twins, have different fingerprints. It has to be admitted, though, that some pairs of identical twins have fingerprints so close that an inexperienced fingerprint technician could easily confuse them, especially if a smudged fingerprint was being investigated. This has a very significant implication – the method by which fingerprints are inherited is unknown; the separation of genes and environment, sometimes called nature versus nurture, is currently impossible in all but a few circumstances to quantify. That we all have individual fingerprints is certain, what is also certain is that there are not an infinite number of different fingerprints.

This is a reflection of the misunderstanding of the word 'infinite'. It does not mean very large, it does not even mean unimaginably large; it stands for a concept of an amount that has no end. For example, the phrase 'the infinite variety of life' is often used and hardly ever questioned. There is simply no such thing as an *infinite* variety of life. We often confuse very large with infinite. Consider it this way: a human has approximately 3,200,000,000 (3.2 billion) DNA bases in every cell, so we can say that the number of possible arrangements of these bases is 3.2 billion factorial. Now, the mathematical idea of a factorial can be simply explained. 5 factorial (normally expressed as 5!) is 5 x 4 x 3 x 2 x 1, which comes out as 120. If you try this with a calculator that has a factorial key you will find that the maximum number that can be factored is about 70, and unlike other calculations it takes a surprising time for the machine to produce an answer. This is because it is multiplying with huge numbers.

The point of this is that 3.2 billion factorial gives a truly immense number, but it is not infinite. Similarly, if there are a finite number of atoms in the universe, it does not make any difference how many there are, the number of possible permutations is not infinite. However, as ever, there is a caveat. All of this meandering around on the nature of infinite is about the following: the permutations of fingerprints are not infinite; however, with an idea of the size of the numbers involved now, the number of individuals sharing a fingerprint at any one time is very small. Indeed, the number of potential fingerprints is so large that it is possible that no two individuals have ever shared the same fingerprint at the same time.

We have found our unique identifying mark then, but here lies a problem – interpretation of a fingerprint. How do you categorize and catalogue a mark in front of you? There need to be rules of interpretation, but these are often very flexible. With a system that has the power to put people away for life its rules of interpretation need to be fixed and inviolate.

If you look at your own fingerprints you will notice that there are various patterns visible, which vary from finger to finger. The three main fingerprint types are shown in figure 1, as are some of the features that are used in analysis. These features are not exhaustive, but serve to indicate some of the variation that can be found. It may be apparent from these diagrams that there is not a great deal of variation, but fingers may have all manner of ridge patterns, which may be combinations and variations of these three basic patterns. The three fingerprint types tend to be represented in the population in stable frequencies, so loops occur 70% of the time, whorls 25% and loops on 5% of fingers.

Fingerprints for identification in the UK had a rather unusual start. A Scottish physician, by the name of Henry Faulds, was very interested in fingerprints and while working in Tokyo during the 1870s amassed a large collection. Such was his expertise that he was asked by the authorities to investigate a crime where a sooty fingerprint had been left on a wall. By comparing the print with the fingerprints of the suspect he was able to show that it was not the man. A second suspect did have a matching fingerprint. This set Faulds thinking and after another successful identification of a criminal from a crime scene fingerprint, he was convinced that fingerprints were a perfect way of identifying an individual. He was so convinced about this that he published his results in the journal *Nature* in 1880.

The process of analysing and comparing fingerprints was now possible, but what was needed was a reliable

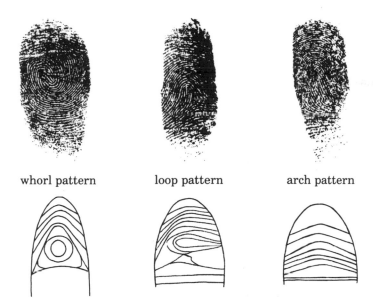

whorl pattern   loop pattern   arch pattern

Fig. 1 Fingerprints and fingerprint features

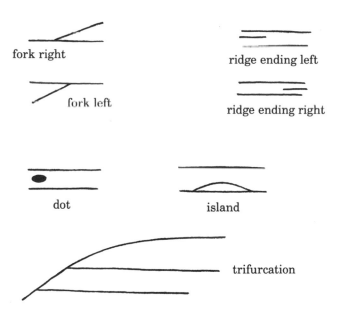

fork right

ridge ending left

fork left

ridge ending right

dot   island

trifurcation

Fig. 2 Fingerprint features used in identification

system of classification. It was Sir Francis Galton who produced the method to do this: it is still the basis of the system of fingerprint classification that is in use today. Galton was a very versatile thinker and came up with a number of innovations, such as the first weather map, which was published in *The Times* in 1875. In 1892, Galton published *Finger Prints,* which laid out the basic methodology of analysis and a system of classification.

In detection of crime, we look for features to match to a profile, or features to search for in the general populace. However, just as fingerprints were used to guarantee cheques, they also figure in predictions of the future, especially in Chinese and Indian folklore. In the Chinese analysis of fingerprints, if you were to have a whorl on a single finger it would indicate poverty; whorls on two fingers conversely indicates riches. It can get even more extreme – whorls on six fingers indicate a thief. The Hindu version of this has a different interpretation on finger patterns. For example, a loop on one finger means the subject is happy, but as the loop number increases, from one finger to five fingers the indications become progressively worse. All of a sudden when the number of fingers with loops goes beyond five, the auspices become progressively better and better, until with loops on nine fingers you must live like a king.

Western science might dismiss this; it isn't this book's role to give an opinion. However, the astrological use of *patterns* can give fingerprint science a new area to search in. There are potentially interesting things that we can definitely determine from fingerprints and palm prints and the patterns therein. For example, it is impossible to determine family relationships from fingerprints, but it is possible to determine large-scale genetic problems from finger and palm prints. It would appear that the finger and palm patterns are very closely related to severe chromosomal abnormalities. So, for example, there is a strong tendency in children with Down's syndrome for the fingers to have loops rather than arches or whorls.

The way in which fingerprint analysis works is quite instructive. In an ideal situation it would be possible to compare a complete fingerprint from a scene of crime with an individual that was under suspicion. Unfortunately this is nearly always impossible. Most countries retain fingerprint collections of known individuals in the hope that a match with a fingerprint from a crime will reduce the number of suspects which have to be looked for, but comparison is made not using the whole print, only sections of it.

Fingerprints are normally classified into basic patterns, as shown in figure 1, including combinations of the patterns and scars, after this the ridge count is used. By measuring features relative to each other a numerical picture can be produced which can be used for comparison with other prints. This sounds straightforward, but it is fraught with problems. A simple example is the case where a smudged print could either erase, or add in, a dot or ridge ending. Because of this it is normal for the discretion of the fingerprint technician to be used to declare a match between fingerprints, which introduces a subjective element into what most people imagine to be an objective procedure. The belief that fingerprint analysis is infallible has grown over the years since it was first used, but by allowing the opinion of the fingerprint technician to decide whether two match or not means that fingerprint evidence is open to dispute. In fact it has been said that if this knowledge were introduced into court now the situation would be quite different. With our increasingly well-educated and vocal population it is conceivable that the sort of arguments about the use of DNA evidence in court that happened on its introduction would also have hit fingerprint evidence. There is little doubt that the standards of accuracy and reliability of fingerprint matches would be rigorously contested.

It is obvious from this that although fingerprints can be used to identify individuals and sometimes associate them with a scene of crime, there is one aspect of personal identification which is of great interest, one where fingerprints cannot help. This is in paternity disputes. Fingerprints are so disparate in their expression that there is no apparent association between the prints and family relationships.

We can see a slight connection between the genetic composition of an individual and their fingerprint pattern. This got scientists thinking, and we get much closer to the genetics of an individual by looking at various proteins. Proteins are the direct product of genes, but as we shall see, this is still one step away from direct analysis of DNA, and consequently less informative.

The earliest analysis of a protein for personal identification was of the ABO blood groups. These were introduced by Carl Landsteiner when he classified bloods of human beings into different types, the now well known A, B, AB and O. For this work he received a Nobel Prize in 1930. Blood groups are interesting because they do not occur at the same frequency. Blood groups A and O have about the same frequency at 42% and 47% respectively, while group B is found in only 8% of people and AB in only 3% of individuals. This means that it becomes possible to potentially match a bloodstain to an individual. Of course, the discriminatory power of such a system is quite low, but there is a use in which blood groups are as powerful as any method and that is in *excluding* a person. So although a bloodstain might match an individual if it was group B you could only say that you would expect such a match to be found in about 1 in 12 individuals. On the other hand if the blood groups did not match, say the stain was B and the individual O, then it can be said with absolute certainty that the blood did not come from that person. The ability to exclude an individual is the most powerful aspect of this test.

An area where blood groups have been used successfully is in paternity disputes. For this to be used, however, we have to know something about their inheritance. Every cell in your body has two copies of every gene, one inherited from your mother and one from your father, with the exception of the genes on the sex chromosomes in men. So each expressed blood group (called the phenotype) is created by two genes, the genotype, as shown below.

PHENOTYPE	GENOTYPE
AB	AB
A	AA *or* AO
B	BB *or* BO
O	OO

You can easily see by thinking backwards that for a child to have a blood group of O the parents would have to have genotypes of AO, BO or OO, so if the suggested father is blood group AB, his genotype must be AB and therefore he could not possibly be the father. The converse is not true: just because the suggested father of an O group child is also O, we cannot say that he definitely must be the father. After all, the father could be any one of the 23.5% of the males in the population that are also blood group O. One method that can be used to improve the discriminatory power of blood groups is to test other groups and multiply the probabilities together. So, for example, if you take the probability of an individual being group O and the probability of also being rhesus negative and multiply them together you get the probability of a person being O negative. Add yet more blood groups in and the probability of finding an individual of that type becomes progressively smaller. It should be remembered though that no matter how many blood groups are added in, the probability, however small, is still finite, so although someone can be completely excluded with 100% certainty, the same certainty cannot be made about inclusion.

This had unfortunate ramifications for the well-known comic actor Charlie Chaplin. In 1943 an actress by the name of Joan Barry filed a paternity suit against Charlie Chaplin. Her claim was based on Chaplin being the father of her unborn child that was due in October of that year. Chaplin was also charged with 'white slavery' and violating the human rights of Barry. This was a slang term for the 1910 Mann Act, which prohibited the transport of women across the state line for 'immoral purposes' and was originally designed to stop prostitution. Unfortunately for Chaplin the case was heard in the California state court system. It was unfortunate because at that time blood group evidence was not admissible in Californian courts. The first trial ended as a mistrial but the second resulted in Chaplin being declared the father, contrary to the blood group results, so Chaplin was ordered to pay child maintenance until the child was 18 years old. The evidence that the blood test results provided was ignored.

Although there are many different blood groups that can be added into the equation, these have never been widely used in forensic applications, being of greater importance in tissue compatibility and organ transplantation. It is really only with the advent of DNA analysis that reliability of personal identification becomes possible. In fact, as we shall see, the range of applications for DNA profiling is becoming greater all the time and the sensitivity of the test is now at its ultimate resolution – producing a DNA profile from a single cell.

2 WHAT IS DNA AND HOW DOES IT WORK?

The most universal molecule associated with life is deoxyribonucleic acid, otherwise known as DNA. In fact, DNA is quite fundamental to life. How we came to know that DNA was pivotal in all life on this planet is quite a story.

The twentieth century has seen the rise of genetics to the point now where the twenty-first century is to be the century of biology. Chemistry and physics have had their landmark times in previous centuries; now it is finally the time when biology takes centre stage. Those of us that have had the good fortune to watch the science of genetics evolving are truly privileged to have been a part of it. However pioneering we might be, however, we would do well to remember the words of Isaac Newton, in a letter to a colleague, Robert Hooke, dated 5 February 1675: 'If I have seen further it is by standing on the shoulders of giants'. We all stand on the shoulders of giants when we undertake scientific investigations.

From the very earliest times of sentience, people have wondered about the way in which characteristics are passed from one generation to another, but it had always been guesswork, often based on incorrect premises and inadequate observations. As time went by, science got closer to the secrets. By the end of the nineteenth century two true giants had published their work, which laid the foundation of genetics. These were Charles Darwin and Gregor Mendel. Even though we now see their work as extremely important, these two careful scientists were at the time simply producing functional descriptions of observations they had made. There was no attempt to explain how characteristics were inherited, or in the case of Darwin, what the precise mechanism of evolution was.

Evolution is not often thought of as part of genetics, but it is all about inheritance and that is a large part of what genetics is. Our knowledge of evolution wasn't always at the fairly enlightened level it is now. It was always obvious that horses only gave birth to horses and cows only produced cows; it was so self-evident that the question never arose as to why it was like that. The assumption was that it just 'was'. It was also assumed that species were immutable. The church held this view and also the huge periods of time which evolution requires is beyond any easy comprehension. One of the most defining ways in which Church control was expressed was with Archbishop Ussher declaring in the 1650s that the world was created on Sunday 23 October 4004 BC. This was expressed in *Annales Veteris Testamenti*, published in 1650. It was a calculation based on working back through the genealogy given in the Old Testament and counting up the generations. One of the problems lies with the time-scale for evolution. It is thought that 'life' began approximately 3.5 billion years ago. Now, consider a life span of, say, 80 years; this figure encapsulates an individual's memory. A social memory might be 1000 years, before the mists of time start distorting the past. So trying to comprehend 3.5 million times that is going to be extremely difficult. It would have been easier, then, to go with the rather more human scale of a few thousand years of history put forward by the Church.

Another difficulty with the acceptance of evolution was realized by Darwin himself. There was no clear and testable hypothesis as to what the mechanism that drove evolution was. Nor was there any clear idea of how a 'characteristic' was inherited. For most people it was a simple case of blending characteristics together: this was quite a reasonable assumption because of very simple observations of hair colour and skin colour. In fact Darwin himself assumed that characteristics blended, which meant that to have a plausible explanation of the workings of evolution it was necessary to assume that an individual with an advantageous characteristic would have to mate with an individual with the same advantageous characteristic, or the advantage would become blended and therefore lost. In some cases, single genes can still appear to blend. For example, if you cross a red snapdragon with a white snapdragon, all the resulting plants will have pink flowers, but now, if the pink flowered plants are cross pollinated the result is some pink flowers, some white and some red. Without having some idea of the mode of inheritance, results such as this are very difficult to interpret.

Into this situation came the other great innovative thinker: Gregor Mendel. He was the son of a peasant farmer, and was christened Johan. It was when he entered an Augustinian monastery at the age of 21 in 1843, that he changed his name to Gregor. Between 1856 and his election as Abbot in 1868, Mendel carried out meticulous research which showed immense attention to detail. When Mendel published his results in 1866 they were virtually ignored. The arena in which he read his work out and published in lacked prestige: he read them out at the Brunn Natural History Society, and then published them in their journal, in German. At the very same society he also read out the meteorological data for the area.

His results in plant breeding laid more or less unknown and undiscovered until about 1900 when several scientists independently rediscovered the work. When the full importance of the work was understood the paper was translated from German into English and republished by the Royal Horticultural Society in England. What Mendel had managed to show was that characteristics were inherited as discrete units, which are either dominant or recessive. Nowadays we know the situation is far more complicated than this simple statement, but when Mendel made his great strides forward in suggesting a mechanism of inheritance the examples he chose were not only fortuitous, but also clear and easy to understand.

When we think of dominant and recessive characteristics we are really only dealing with a very few particular traits that are controlled by a single gene. One such is tongue rolling in humans – a single gene controls the innate ability. If a gene controls a metabolic pathway and makes a defective product, it is often recessive, since it affords no advantage to the organism. This rule of thumb for recessive and dominant does not always stand true. There are cases of rare disorders where the disease is inherited in a dominant fashion, such as Huntington's chorea, a degenerative neurological disease. Now it is known that many conditions and inherited traits are controlled by large numbers of genes, often with an environmental influence as well.

When Mendel carried out his work he very carefully hand-pollinated peas. These were ideal plants for what he had in mind because they usually undergo self-fertilization. Since the anthers (which produce pollen) and the stamens are completely enclosed this also stops accidental pollination by insects. As well as this, pollen production takes place before the stigma that receives the pollen, so if you were to remove the anthers it is possible to closely control the cross-pollination mechanism. Mendel had spotted that there were certain characteristics that were inherited as discrete units. These included the appearance of the seed, which was round and smooth or wrinkled, and whether the stems were long or short.

With this work it became possible to see how it might posit a very real chance of explaining the basic mechanism of inheritance and tie in other observations about cells which had been made earlier in the same century. Cells had been investigated over many years using the relatively new techniques of microscopy, which were slowly becoming available. It was during this period that *chromosomes* were first recorded as having been seen.

Chromosomes are complicated constructions of DNA and proteins and they carry the entire DNA, which forms the genetic material. Chromosomes are carried in the cell nucleus, and in humans each cell normally has 46 chromosomes. This figure was not actually arrived at until the middle of the twentieth century, but an American scientist, Thomas Hunt Morgan, firmly established chromosomes as the basis of heredity in 1907. At this time, although their importance was known and also that they were broadly made up of DNA and protein, what was definitely not known was whether it was the DNA or protein that actually carried the genetic information. So, armed with the information available, it became possible in the following years to start mapping the position of genes along chromosomes, producing a functional description of genes and their influence, even though what they were made of was unknown. During the following 40 years the debate raged as to what the chemical nature of the genetic material was. During this period a great deal was discovered about genes, such as that they can be mutated by X-rays and various chemicals.

Developing a clear idea of how inheritance and evolution interact is tricky and there is one major case that eventually had devastating results for an entire country. Early in the nineteenth century a French naturalist, Jean Lamarck, produced a pre-Darwinian theory of evolution. Broadly speaking his thesis was that characteristics acquired in one generation could be passed on to further generations. He was not suggesting that mutations in DNA were passed down the generations: DNA and mutations were completely unknown at this time. His most famous example is the neck of the giraffe. He suggested that if in one generation a giraffe stretches to reach food high on the tree, the ones with longer necks have the advantage regarding food. The next generation have longer necks and so the long neck of the giraffe developed. Now, this innocent example would not have been of any consequence except that it so well encapsulated his theory that it was taken up by a revolutionary Russian called Trofim Denisovich Lysenko (1898–1976).

Lysenko was born in the Ukraine and by 1930s had gained quite a reputation as an agronomist. What he did was effectively ignore Mendel's work on inheritance and propound the idea that if crops are grown in poor conditions the seed from succeeding generations will be adapted to such conditions and therefore prosper. Unfortunately this is not so, yet Lysenko became the director of the Institute of Genetics of the Soviet Academy of Sciences from 1940 to 1965 and not only did he declare Mendel to be wrong, but he suppressed any Soviet geneticists that opposed him. The result, in practical terms, was the cultivation of inappropriate crops in inappropriate areas on collective farms, which resulted in famine in what was then the USSR. This was based on the idea that if you grow a good crop in drought conditions the ability to survive will be passed on to the next generation: effectively they 'learn' and the learning is passed down. This does not happen, but as a consequence of this dogma there was widespread crop failure and the science of genetics was set back in the USSR. More importantly, people starved due to inappropriate and badly thought through theorizing.

To try and work out in a logical way, with no experimental data to go on, whether it was DNA or protein that was the material that made up our genes was going to be extremely difficult. The argument went along these lines: we are very complicated, so something as chemically simple as DNA could not possibly carry the necessary information to make us what we are; on the other hand proteins are chemically quite complex, so they must be the genetic material.

It was only in 1944 that three American scientists carried out a series of very elegant experiments, which finally confirmed that it *was* this very simple molecule, DNA, which is the genetic material – the very stuff of our genes.

As we know, DNA is the fundamental building block of life. It is effectively a digital information code for all of our genes, but just knowing the chemistry of this molecule does not fully explain why we are what we are. Conversely, if we did not know the chemistry of DNA we would know absolutely nothing about how we inherit the features that make us individuals and we would not be able to use DNA as the most powerful tool in personal identification, which it most definitely is.

Essentially, DNA is a molecule made up of just four

elements, carbon, hydrogen, oxygen and nitrogen. These
elements are put together in different ways to form four
different structures: the bases. As you can see in Figure 3,
the bases are all different, but only slightly. Stringing
these bases together using a common backbone results in
a DNA helix. The four bases are called adenine (A),
guanine (G), thymine (T) and cytosine (C). Adenine and
guanine are referred to chemically as *purines*, while
thymine and cytosine are *pyrimidines*. The chemical
difference between these two is small and subtle but basically the purines are larger than the pyrimidines.

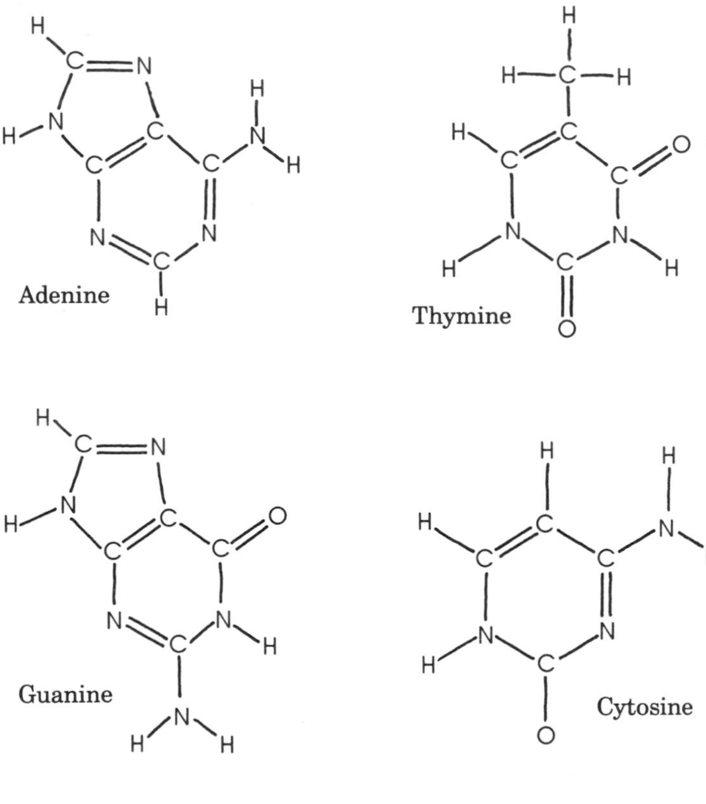

Fig. 3 The four bases that make up the genetic code

Now, because they are different shapes, a purine will bind with a pyrimidine; however, it is more specific than that. Thymine will only bind with adenine and guanine will only bind with cytosine. So if you have a thread of bases attached to their backbone of sugars, effectively a single strand, it is possible to re-create the entire double helix by simply adding the relevant matching bases. This is a fundamental aspect and the power of DNA. It is a double helix and can be reconstituted, in its entirety, from only half the original double strand. This is the very way in which cells replicate their DNA: the double helix divides and each half forms the template for two completely new double helices, so DNA can replicate itself to give the continuity of life. It is part of what makes DNA so amazing.

The uniformity of genetic coding across the entire gamut of both plants and animals is astonishing. Although DNA is the molecule that codes for the entire blueprint of life, life is not that simple. We are all, even identical twins with the same DNA, slightly different. This reflects the complicated interactions found between genes, gene products and the environment. Just as the genes code for all of the proteins, which go to make up you and me, they also give you your potential, but it is just that – potential. For example, no great athlete was ever born fully-formed, to be the best. No matter what their potential, an athlete has to train to reach their full potential. So if we were given a DNA sequence, could we tell if it was an athlete or musician? The answer to that is no, but perhaps surprisingly even more fundamental questions are nearly as difficult to answer just using DNA. One such question is has this DNA come from a plant or animal? Unless the sequence is very well characterized, for example, so that you know what the sequence represents, this is extraordinarily difficult to determine. As the evolutionary distance between the two organisms which you are trying to identify gets smaller, determining them just from a random sequence becomes ever more difficult. All this shows us that all plants and animals are related and using DNA to identify an individual from among a group of people has to be carried out very carefully, or mistakes will happen.

Whenever a DNA sequence is used for identification it has to be thoroughly characterized and very often the sequences that are used do not actually code for anything. Because these sequences do not code for anything, they are free to evolve and change without affecting the individual. Examining some of these non-coding DNA sequences reveals that even if you look at just one of them, when a group of different people are investigated, the sequence varies between them in a measurable way and since it is a measurable way it is possible to say that the DNA is different between two people.

To make a DNA profile there are two broad types of DNA that can be used, there being two types of DNA present in every cell, each of which has its own use, both of which are complementary to each other. These two types of DNA are nuclear DNA, found only in the cell nucleus, and mitochondrial DNA (mtDNA).

Mitochondrial DNA is unusual in that it is generally inherited only from your mother, which, as we shall see later, can be used to our advantage. This, of course, needs an explanation. It sounds simple but we have to go back in evolutionary time to understand why this happens. Mitochondria are the powerhouses of the cell; they are the place where energy is generated for every activity that cells, tissues, muscles and ultimately your entire body wishes to perform. But we did not always have mitochondria. When the world was young and life started, some 3,500 million years or so ago, the single cells that started out were just bags of proteins, enzymes and, of course, nucleic acid.

In evolutionary terms the problem with this is that it is extremely difficult to produce a large multi-cellular organism without some sort of internal structure to cells, so that cells can differentiate into different tissues and organs. So at some time in the far distant past a fortuitous event took place with the invasion of a primitive cell with an equally primitive smaller organism. That this had happened was only realized when it was found that mitochondria contained DNA which was independent from the DNA found in the nucleus. This mtDNA also had one of the bases slightly altered. It was different from the DNA found in the nucleus. This, all of a sudden, matched up with a long known observation, that under the microscope mitochondria can be seen to elongate and divide like bacteria. This process of division was also known to be independent of the division of the cell. This is important because mitochondria are the powerhouse of the cell and therefore the body as a whole: it is these organelles that actually produce energy for use in metabolism. So some cells that have a high metabolic rate have a much larger number of mitochondria than cells that have a much lower basal rate. A form of fatty tissue that is found mainly in babies and the young of many mammals is called brown fat. This brown fat generates heat in newborn mammals and is the reason that they do not shiver. The brown fat manages to generate enough heat to keep the child warm and the reason it is brown is the high concentration of iron-containing enzymes that are associated with the mitochondria that generates the heat.

The reason that mitochondria are so important in personal identification is two-fold. The first of these is that it is possible to track a maternal line of descent and the second is the very robust nature of mitochondria. This robustness is mostly a function of the sites at which mito-chondria are found. While nuclear DNA is mostly confined to soft tissues, mitochondria can be found embedded in bone and teeth as well as hair shafts. These are very resilient to decay, so in general the last nucleic acid to disappear is mitochondrial DNA (mtDNA).

When an egg is fertilized by a sperm it generally supplies no mtDNA to the egg. The sperm in fact does have a mitochondrion, a single entity, which does not generally enter the egg upon fertilization. The sperm needs this mitochondrion to provide the energy for the flagellum, or tail, which propels it along.

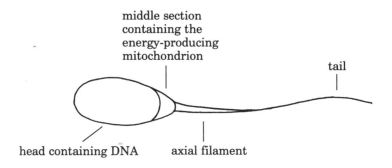

middle section containing the energy-producing mitochondrion

tail

head containing DNA   axial filament

Fig. 4 Schematic representation of a human sperm
Shows the large single mitochondrion which generates energy for the swimming action

So, as the foetus grows, only the mitochondria from the mother are present in all of the cells, whether the child is either a girl or boy. Knowing this has been of immense value in some cases.

So, we have, broadly speaking, two different types of DNA, nuclear and mitochondrial, both of which can be used to make a DNA profile, but just how is this done? There are several methods available and they have all been used at various times.

Alec Jefferys, who was working at Leicester University, devised the first method in the 1980s. It was not a project deliberately directed at forensic applications, it was a project that was designed to track down disease genes. This method produces what looks much like a bar code when the process is finally finished. What was quickly spotted was that the patterns of these bars between individuals were significantly different. Professor Jefferys developed two different techniques for developing DNA profiles. These are referred to as Single Locus Probes (SLP) and Multilocus Probes (MLP). Both of these techniques pick up small differences in a very large amount of DNA. Each human cell contains approximately 3000 million bases, wrapped up in 46 different chromosomes of different sizes. Most of this material is more or less the same between individuals, since we all need the same genes for growth and metabolism, but amongst all this material there is some variation, so the differences that are looked for are tiny. Importantly, the variation in an individual is the same in every cell in the body, so irrespective of what the source of the testing material is, the results will always be the same for that person.

These two different profiling techniques, MLP and SLP, are both produced in the same way. It is only the end result which is different, and very different they are too. The technique is broadly carried out in five steps. The first is to extract DNA from a blood sample. This is a well-characterized process and can in fact be carried out in a domestic kitchen using domestic chemicals and utensils, although such a recipe would need to be followed quite carefully. Once extracted, the large pieces of DNA, one from each chromosome, are cut up into different lengths. This is done using enzymes called endonucleases. These are all generated from bacteria of various sorts. If any given enzyme is used it will always cut the DNA in the same places. So, if a cocktail of an individual's DNA from millions of cells is treated with an enzyme, it will always produce the same number of same-sized fragments.

The next stage is to run out these various sized DNA fragments on an agarose gel. Agarose is based on agar, which is an extract of various types of seaweed; originally it came from the Far East where in Malay it is called agar-agar. The gel is a clear material of about the same consistency as a stiff, edible jelly. When an electric current is passed through the ·gel from one end to the other it moves the DNA fragments down the gel, but not all together. What happens is that the small fragments move down the gel faster than the longer ones, consequently the fragments become separated. This process is broadly based on the idea that the longer fragments take longer to find their way through the agar matrix than the shorter ones. Of course at this point the individual fragments cannot be seen and because of the huge size of the human genome, if all the fragments could be seen at the same time the gel would just look like a continuous smear, a solid block of DNA. it is amazing that any information can be extracted from such a gel, but it can be, although only by another step in the long process.

Now, at this point the DNA will simply diffuse passively if left to its own devices, so it has to be fixed so that it can be manipulated further. This is done quite literally by blotting the gel and using the process of passive diffusion, but vertically. A nylon membrane is put on top of the gel, then lots and lots of paper towels; the paper towels absorb the water from the gel and the DNA travels with it. The DNA gets stuck on the nylon membrane and then can be fixed in place.

Once fixed on the membrane the DNA can be probed. There are many different ways of doing this. The original method was using radioactive probes. A probe is a small piece of complementary DNA that will bind to a specific sequence of DNA. If the probe binds to several sites, being a general probe, then it is a multilocus probe. If it only binds to a single site, being highly specific, then it is a single locus probe. Using a radioactive probe, after the probe has become bound to the DNA on the nylon membrane it can be placed against a piece of X-ray film and over time the radioactive probe blackens a small patch on the film. When the film is developed you have before you a DNA profile. This was originally called a DNA fingerprint, but this was

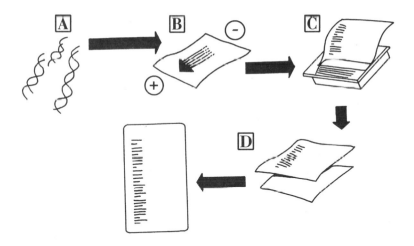

Fig. 5 How to make an SLP or MLP DNA profile

Take a source of DNA and cut it up into small pieces using endonuclease enzymes. Pass the fragments through a gel under the influence of an electric current. The gel is 'fixed' so that no further movement of the DNA is possible. The DNA is transferred to a stable membrane which can then be probed using either radioactive probes or chemiluminescence to produce a DNA profile

thought rather confusing, and created some antipathy amongst traditional finger printers who thought that calling it a DNA fingerprint was hijacking the word.

Although biology laboratories had routinely been using radioactive probes for all manner of investigations for many years, it was quickly appreciated that the explosive growth of molecular biology meant that radioactive probes were going to become unsustainable, simply because of the problems of disposal and risk to all the other members of the laboratory. As a direct consequence of this and the very high commercial value of producing material for molecular biology, companies started producing alternatives to radioactive probes.

The favourite alternative to radioactive probes was to include a molecule that could be activated to give off light. This *chemiluminescence*, as it is called, did use toxic chemicals, but nowhere near as persistent as radioactive material. It also had the advantage that it was quicker than radioactive probes to produce an image, although it still needed an X-ray plate to create a permanent image. Alongside the material that is being tested the only way of finding out whether the process has worked properly is to run out standards at the same time. These are fragments of DNA of known size which react and bind to the probe being used to produce a molecular weight ladder so that it is possible to tell, more or less, not only which bands match, but the size of the matching fragments as well in terms of the number of bases present in the fragment.

The reason that the number of bands varies between MLP and SLP profiles is that MLP bind to lots of different sites among all the broken bits of DNA on the gel. SLP systems only recognize a single sequence of DNA. Although the probes, of either type, only recognize a specific sequence, the overall length of the sequence may vary because the enzyme cutting sites may not be present in all the same places. Consequently the length of the fragments can be quite different, resulting in an individual pattern for an individual person.

The number of bands of both of these methods can also vary between individuals. This variation is because all higher organisms have two chromosomes, one from the mother and one from the father. With SLP analysis, if both of the bands are the same size it is called homozygous and although there are actually two bands present, only one can be seen. If the sequence length is different between the mother and father the result will be two bands and this is called heterozygous. Although rather more difficult to interpret because of the large number of bands present, the same holds true for MLPs.

Both MLP and SLP profiles sound very straightforward, but the bands on the final image are not always clear and sharp edged. This resulted in forensic science service laboratories having to agree on exactly how these profiles should be interpreted. It also meant that it was virtually impossible for any two laboratories to compare results between themselves, especially laboratories in different countries. Another result of this was that for all the hype associated with these systems it was going to prove impossible to produce anything resembling a useable DNA database, which was later going to become the aim of all forensic laboratories.

With the complicated and slightly subjective interpretation of these two systems the search was on for something reliable, as statistically discriminatory, and simple enough to allow the establishment of a national database and even allow international comparisons. Another aspect of this was that if a large database were going to be produced a large number of staff would be required, so a simple, automated system would be best. Such a system would require less training for staff: in effect they would not need to understand the process, just follow a standard set of instructions. It took a while, but an answer was eventually found. This method, which is currently used all around the world, is called short tandem repeat (STR) analysis. STR analysis works in a quite different way to MLP and SLP techniques. It is so different in fact that the profiles generated by MLP and SLP technology are literally impossible to compare with STR profiles.

It is not only the final profile from STR analysis that is quite different, the methods used to construct the profile are also quite different. It all started with an American, Kary Mullis, who devised a new technique, for which he was awarded a Nobel Prize in 1993. This technique is called the polymerase chain reaction (PCR).

PCR works by using naturally found enzymes, which are used in cells to replicate DNA. By targeting specific sequences and going through several cycles of PCR it is possible to replicate your target sequence exponentially. So if you start with DNA from a single cell (this is possible) which contains two copies of the target DNA, one from the mother and one from the father, the first round of PCR produces 4 target molecules, the two original and the two copies, the next round of PCR 8, then 16, 32, 64, 128, 256, 512, 1024 and then after ten cycles, 2048. Typically there would be more than twenty rounds of PCR, so from a vanishingly small starting point large amounts of target DNA can be produced, large enough amounts, in fact, to be analysed.

This DNA created by PCR can be analysed using STR techniques. It is important that the target for the PCR contains an STR. STRs are scattered throughout the entire genome; the precise number is not known but they make up a considerable amount of the total DNA. An STR is a short sequence of DNA which is made up of simple repeats, which may take the form of 1, 2, 3, 4, 5, or 6 bases long. So a simple STR might be – ATGATGATGATGATG or – CTTTCTTTCTTTCTTTCTTT.

As you can see these are short repeat sequences that are repeated several times and it is the number of repeats that is important in STR analysis. The number of repeats is variable between individuals, although for most of these STRs there is generally a relatively small amount of variation in the number of repeats for any given STR in a population. There are exceptions to this but they tend to be associated with clinical conditions. These repeat sequences were once referred to as junk DNA, but we now know that although highly variable they are important in the control of gene expression.

There are two copies of every chromosome, with the exception of males, where there are two quite different sex chromosomes, X and Y, females having two X chromosomes. Half of these chromosomes come from the father and the other set comes from the mother. Because of this there are two copies of every STR in every cell. Although the copies will have the same sequence of bases, they may have different numbers of repeats. If the repeat number is the same on both chromosomes they are described as homozygous and there will only be one band; if, on the other hand, the repeat number is different on the two chromosomes they are described as heterozygous and there will be two bands.

At this point the process becomes highly automated. The STR is loaded on a gel, again with a comparison lane containing a known reference sample. As they have all been tagged with a fluorescent dye, as they move through the gel the dye is fluoresced by a laser and a time-resolved image produced which gives the relative positions of all the bands. The relative position of the bands depends on how many repeats are present, the greater the number the slower it proceeds. So the time it takes for a band to travel from the top to the bottom of the gel reflects the number of the short tandem repeats which are present. Consequently, when compared with the length of the known controls, it becomes possible to assign a simple number to the STR, which directly reflects the number of repeats present.

This means that when given the STR equivalent instead of a multilocus or single locus probe DNA profile, the STR profile is a series of numbers, instead of an image which has to be controversially compared by eye. No images, no subjective comparisons, no possible/possibly-not arguments, just a straightforward number which has the advantage that with adequate standardization of techniques, two laboratories on opposite sides of the world will produce the same stream of numbers.

The story does not of course end there and the next step depends upon whether the profile is to be used in a criminal case, such as rape or murder, or whether it is in a civil case such as a paternity dispute. Either way a probability of the profile turning up by chance has to be calculated. This is done in an interesting way. For every STR that is used a whole range of samples from the general population are needed to generate a probability curve. This is the first essential of a DNA analysis because not all of the different sizes of STR occur with the same frequency throughout the population; in fact some of the currently used STRs have some sizes that turn up in 80% of the population. These are therefore described as having very low discriminatory power. So once there is a good, representative, sample from the population it becomes possible to be able to say what the probability is of finding any given STR by chance alone among a group of unrelated individuals. It is important at this stage to be sure that they are unrelated, the reason for which will become apparent later. If the result of measuring a single STR in an individual produces, say, a value of 14, 16 (that is there are 14 repeats in the STR of one chromosome and 16 on the other chromosome), it is possible to go to the comparison database and find that this repeat number is found in 20% of individuals. To put it another way, one in five people will have that value of 14,16. This is not a very good figure. Imagine going into court as the prosecution, only to say that the DNA analysis indicates that there is a 1 in 5 chance that the DNA came from the defendant, only for the defence council to point out that since there were 12 jurors, a judge and two barristers, 15 in all, it would be reasonable to suggest that there would be three people in court with the same value of 14,16. What we need is a way to improve the discriminatory power. This is surprisingly easy.

Since if they are carefully chosen, STRs are unrelated to each other, a simple multiplication can be used to generate a new value for the probability of finding a particular combination of STRs. This is easily shown by an example. Our first STR has a value of 14,16 and is found in 1 in 5 individuals, the second STR from the same individual has 20 and 25 repeats on each of the two chromosomes and is found in 1 in 4 of individuals, that is in 25% of the population. If we change these to simple proportions we have the first STR with a probability of 0.20 and the second STR with a probability of 0.25. If we multiply them together, we get 0.05. This is the same as saying 1 in 20 individuals would be expected to have the different STRs of 14,16 and 20,25 in that combination. All of a sudden we can start to generate greater probabilities than from single STRs. This is the basis of modern DNA profiling. With increasing numbers of STRs used in producing a DNA profile, the probability of finding any given profile becomes progressively less and less likely, by the simple expedient of being able to multiply the probabilities together. This is perfectly legitimate and as can be readily understood, by describing the frequencies as fractions it is obvious that when multiplied together the values actually become smaller – a half of a half is a quarter.

When STR profiles were first introduced in the early 1990s only four different STRs were used, so the frequencies that were generated were really quite low by comparison with the current use of ten STRs in the UK, although some countries use more. When large numbers of STRs are measured the possibility that two people will share the same overall profile becomes progressively smaller as the number increases because the probability that an individual has the same collection of STR results as any other individual becomes a matter of simply multiplying probabilities together; this is the DNA profile that says that the probability that this individual shares the same profile with a randomly picked person from the population is one in 10,000,000, or more. So where does this leave us? Well, the one thing that should always be remembered is that no matter what the forensic science service tells us, DNA profiles are not infallible guides to a person's guilt. Even if a match is found between the DNA profile of, say, a rapist and an arrested individual, the question remains 'what is the possibility that this profile matches the profile of the arrested person by chance?'

Using a comparison with the data accumulated from the general population it might be possible to say that it is 100,000 times less likely that this DNA profile originated from another person unrelated to the accused. Note here, unrelated is again used. This is because we do not need a database to calculate the probability of two siblings sharing the same DNA profile, we can work it out on a purely theoretical basis, which gives us the upper and lower bounds of probability.

If we have two siblings, two brothers, two sisters or a brother and sister, a question we can ask is 'what is the probability that they will have the same STR DNA profile?' With this question we can make a straightforward calculation, which will give us a maximum probability of this happening. It is a logical calculation and we do not need to know anything about the STRs being measured. Taking the case of only one STR, there are two versions, one from the mother and one from the father. The parents might be homozygous, that is both copies in the parents are the same, say, 14,14 in the mother and 12,12 in the father. All the children will then have to be 12,14. If the parents are heterozygous, that is the two versions of the STR are different, say 12,13 and 9,15, then the children could be 9,12, or 9,13, or 12,15, or 13,15. There is a one in four chance of the siblings having the same STR profile with one STR being used. As the number of STRs being looked at increases, so the probability of sharing a profile becomes less and less. By the time four STRs are measured the maximum probability of two siblings sharing exactly the same profile, assuming heterozygosity at every STR with the two parents not having any STR values being shared, becomes 1 in 256. If any of the STR values are the same between the two parents, which is not uncommon, this value drops.

Although this works for siblings, such easy calculations cannot be used for unrelated individuals; in these cases access to a database is needed. There are broadly two types of database. The first, as we have already described, is an anonymous collection of data which gives us the frequency of the different sizes of STR in the general population. The second is becoming ever more useful: it is a database of named individuals and samples from scenes of crimes. This database is constructed from DNA samples taken from named persons who have been charged with an arrestable offence. When these samples are compared with the national database of crime samples it may be possible to match an individual to a crime sample from the past. Similarly, it may be possible to link a sample from a scene of crime to a named individual who may have been arrested several years earlier, at which time they donated a DNA sample, which is becoming an increasingly common event. A common question is where the DNA sample for use on the DNA database is taken from. Most people still think it is a blood sample, but this is no longer so. The usual process is by taking a swab from inside the cheek: quick and easy and perhaps more importantly it can be done in a police station by an essentially unskilled individual. This is unlike taking blood, which requires a doctor or nurse to do it. Crime scene samples come in a variety of different guises, such as semen in rape cases, blood, sputum and saliva from assaults and burglaries. In fact pretty well any bodily material will render some DNA that can be analysed.

Problems arise with mixed samples where it may not be possible to distinguish anything other than that there is DNA present from more than one individual. To stop accidental sample mixing, the samples, whether a mouth swab from an individual or a sample from a scene of crime, are immediately put into sealed tamper-proof bags as soon as they have been taken. One of the situations in which a mixture is inevitable is rape. It would be expected that there would be sperm present as well as cells from the victim. To separate a mixture of this sort is surprisingly easy; sperm are really quite tough compared with epithelial cells, and so it is possible to treat the mixture such that the epithelial cells disintegrate and the sperm cells can be retained and then their DNA can be extracted. To check that it is the sperm which generates the profile there is a gene sequence that can be looked at, which appears on the X chromosome and the Y chromosome. The sequence on the Y chromosome is inactivated, only the X chromosome has a functioning sequence. The inactivated sequence on the Y chromosome is slightly larger than the sequence on the X chromosome. This gene is called amelogenin and although fully functional on the X chromosome, it does not function on the Y chromosome. So if it is female material there will only be one band, because the two X chromosomes have same sized gene. But if the material comes from a male there will be two bands, one from the X chromosome and one from the Y chromosome. Using amelogenin to determine the sex of a sample can be very useful since if there is a bloodstain at a burglary it helps to determine if the perpetrator is male or female.

A very important thing to remember about DNA profiles is that strictly speaking they cannot give any information about guilt or innocence, only a probability that the DNA came from a stated individual. There is a condition referred to as the Prosecutor's Fallacy. Broadly this is where the question which DNA profiling can answer, that is 'what is the probability that the DNA of the defendant will match the DNA from the crime scene given that the defendant is innocent?' is mixed up with the question 'what is the probability that the defendant is innocent given that the profiles match?' Of course the second question is the one that a court wants an answer to, but it is only the first question that DNA *can* answer.

Another question that DNA cannot answer is how the DNA got there. Of course in the case of rape the presence of sperm is an exception to this, but at the turn of the twenty-first century it has become routine to take DNA profiles from such items as discarded cigarette ends where skin cells from the lips and tongue have stuck to the paper. These are mobile samples and great care must be taken to ensure that the cigarette end has not been moved from an innocent place to an incriminating position. What a DNA profile can say with absolute certainty, and it really is absolute, and cannot be gainsaid, is when two DNA profiles do not match, they cannot have come from the same individual. Under these circumstances the tested individual is completely ruled out as the origin of the crime sample. No probability is required for this, if it is a simple exclusion, no question can be made about the profile. An exclusion is an exclusion.

The other major use for DNA profiles using STR technology is in paternity disputes. They are nearly always paternity disputes rather than maternity disputes because although there is not normally an audience at the conception, there often is in a maternity unit, so there is usually little doubt as to who the mother is. To determine family relationships the profile data is broadly used in the same way as in criminal cases, but there are some differences.

In paternity disputes there are basically two different situations. The first is when it is necessary to decide which of two putative fathers is responsible, that is when a mother does not know for certain which of two partners is the father of the child. This is effectively a closed system; there is no dispute as to the mother, but she might have had more than one partner. As long as all parties agree that one of the two named men is the father it can be sorted out quite easily. By testing the mother and child, the STR component which originates from the mother and appears in the child can be removed from the equation. Now all that remains to be done is test the two putative fathers to see which one provided the other component of the child's STR profile. This sort of analysis is straightforward and generally not contentious, because either of the suggested fathers would be quite happy with the outcome. In fact the one who is ruled out will often be disappointed rather than pleased.

The alternative, when it does become fraught, is when the suggested father does not want to be identified. It is just such situations where the Child Support Agency (CSA) become involved and tempers can run high. The CSA remains determined to make sure that absentee fathers pay maintenance for their children, as is right, although the amount which is demanded is a political hot potato.

In these cases there needs to be a relatively high degree of certainty that the suggested father, when tested and found to be the possible father, can rely on the result as sound. Very often the argument of the man is that it was not him, it was somebody else, an unnamed other person. Well, it might be, so it becomes important that the probabilities become in excess of 99.9% likely that a named individual is the father of a child, rather than an unnamed individual. So we have to go back to a database of STR frequencies and make a calculation based on the child and the father. Strictly speaking it is not absolutely essential that the mother is involved in the calculation, although it is much easier if she is.

One thing that is very necessary in cases such as this is a large panel of STRs. This is because of a very important aspect of biology which is often overlooked – mutation. When an STR profile is produced in a criminal case it represents the STR profile of that individual, but when STR profiles are used in paternity cases there is always the possibility that the process of mutation may make it apparently impossible that a suggested father is, in fact, the father.

This is because the very high turnover of cells during, say, wound healing, would result in a proportionate number of mutations affecting the number of repeats in an STR. So it might be possible, in a theoretical way, to find a mutation within an individual. This is, however, extremely unlikely. The mutation would be swamped by the bulk of cells used in the test. But it is different with a mutation that has taken place in the *sperm* of a putative father, where a single change in an STR repeat number will be passed on to the child in all its cells, with the result that on first glance the child is not the progeny of the father. Great care needs to be exercised here. We have almost come full circle: back to a subjective interpretation of a DNA profile. It will never be perfect and there will always be room for legitimate argument as to what the true meaning of a DNA profile is; but one thing is certain, this is a very powerful tool in the right hands, but the human race as a collective group must always be wary of people bearing powerful tools.

3 THE FIRST USE OF DNA PROFILING IN A CRIMINAL CASE

1987 saw the introduction of a forensic technique that has changed the face of both criminal and civil investigations beyond all recognition. This story, however, started many years earlier and ended with a twist in the tale. This finally justified the actions of the police, but in a most unexpected way.

Narborough is a small village in the East Midlands county of Leicestershire with a railway station next to a level crossing that stops the road traffic when a train is passing through, on its way to Leicester. Until 1983 there had not been any particular reason for it to be brought to public interest. There was a small nature reserve, Narborough bog, home to some of the more interesting flora and fauna of the area. The next village is Enderby and is only a short walk from Narborough. This quiet rural life was all to end on a cold Monday, 21 November 1983. It was on this date that Lynda Mann started her day like any other fifteen year old, going to school and meeting friends, chatting about the weekend. In the evening of that Monday, Lynda went out to baby sit, but as it turned out she was not required that evening after all. Consequently, after she returned home she decided to visit a friend who lived in the next village.

She started to walk towards Enderby, where her friend lived, which is about twenty minutes' walk from Narborough. Her route took her past Carlton Hayes psychiatric hospital, which has long since closed and the site sold on to a financial institution. This would later give rise to many erroneous suspicions and finger pointing, reflecting the suspicion and fear that people always seem to have regarding psychiatric hospitals.

Lynda's mother and stepfather were out that night, at a local public house in the middle of Enderby called The Dog and Gun, a tied house owned by the local brewer, Everards. They returned home past midnight to find that Lynda's older sister, Susan, was still up and waiting because Lynda had not returned. This was distressing, as it would be for any parents finding their fifteen-year-old daughter was out after midnight on a school day, but doubly so for Lynda's family because she was not a wayward girl and could always be relied upon to get home at a reasonable time.

The immediate reaction of Eddie, her stepfather, was to drive around the villages of Narborough and Enderby, to see if he could locate her. In the cold, early hours of the morning he found no sign of Lynda, even though it transpired that he had been within a few feet of her body on his search.

Just after seven o'clock on that frosty Tuesday morning, a hospital porter on his way to Carlton Hayes hospital discovered the body of Lynda Mann. At first he did not believe it was a body. Most people do not expect to see such a thing; the immediate reaction is to believe that it must be something else. Few people have seen a dead body and those that have very rarely seen one out of context, unceremoniously dumped outside. It was, however, evident that this was a body, clothed above the waist, but naked below. The hospital porter attracted the attention of an ambulance driver who was also driving to work at the hospital. Lynda Mann was dead; sexually assaulted, raped and murdered.

By 8.30am the police enquiry into the murder had started. It was obvious from the appearance of the body that this was no accident and no casual mugging. This was a calculated crime. A home office pathologist was called and the body removed. Lynda's stepfather Eddie Eastwood carried out a formal identification. With a decent respect for the family the post-mortem examination was carried out on 23 November. Lynda was 157cm tall and had officially died of strangulation, although there was considerable bruising. It would appear that sexual intercourse had been attempted, but had not taken place. Her jeans seemed to have been forcibly removed as though rape was the intent, but premature ejaculation was suspected as there was dried semen around the vagina, but no tears or bruises consistent with rape.

One of the first unfounded suspicions landed on the doorstep of Carlton Hayes hospital, simply because it was a psychiatric unit. Another unfounded suspicion was that Eddie Eastwood might be the killer. While suspicions flew in many directions the forensic laboratory had been busy with the samples and already produced significant results.

It turned out that the pathologist's statement that sexual intercourse had been attempted was not the complete picture. Swabs taken from inside the vagina indicated that penetration had taken place after all. At this point it was no use thinking about DNA tests, there was no such thing available: older, less discriminatory tests had to be employed. The one of choice when dealing with semen stains was the PGM test. This tests for an enzyme called phosphoglucomutase which is important in glucose metabolism and comes in different forms. It was these different forms that were tested for. Although this is commonly referred to as the PGM test, in biochemistry PGM is an enzyme called phoshoglycerate phosphomutase. The enzyme which was tested is normally referred to as PgluM.

Sometimes it is also possible to determine the blood group of an individual from the antibodies secreted in semen. Some people have a gene called *secretor* that means that they secrete their blood group in their semen and other body fluids, such as saliva. This helped reduce the number of possible suspects, but it was still potentially a very large group because assuming that a passing individual did not perpetrate this crime, it left the total male population of both villages.

It was quickly determined that the rapist was PGM1+, a secretor. Although it was assumed that the rapist also murdered Lynda, the only evidence at this point was about the rapist. It may seem unlikely, and in this case was so, but it is always possible that she had been raped by one man and strangled by another. This particular PGM enzyme designation is only found in about one in ten males, so although not very precise it could be used to exclude an individual from inquiries. As with all such tests exclusion is absolute, an inclusion is not.

It became progressively more complicated for the police as people came forward with stories of having seen men running or walking or behaving strangely in all parts of the village. None of these progressed any further than just leads; few men were found who could be associated with these sightings and all of those that were found were ruled out of the investigation.

During this period at the start of the investigation the

Eastwoods had been trying to regain control of their child's body from the coroner. Although the post-mortem had been carried out with admirable swiftness it now became apparent that until all forensic avenues had been explored this was not going to be possible. Consequently it was not until early February 1984 that burial was permitted by the coroners. The funeral took place in the local Narborough church with a large congregation of local people, many of whom had known Lynda.

Meanwhile the enquiry rumbled on, with no new evidence coming to light, until around Easter of that year when the locally assembled murder squad was essentially disbanded, quietly and discreetly, and sent back to their various units. This did not happen overnight, but as the total amount of work declined so the number of officers carrying out the investigation similarly declined. Investigative tools, such as house-to-house inquiries, became less and less important and other more pressing work took precedence. The final, formal, closure of the investigative team occurred at the end of the summer in 1984.

It was this same year of 1984, when the investigation was wearing down, that Alec Jefferys produced the first useable DNA profiles. At this early stage it was often, in fact usually, referred to as a DNA fingerprint because of the potential for individual identification. This name was soon dropped when it started causing confusion in the minds of police and public alike with the literal finger-print. By the time the information about genetic profiling was in the public domain, which occurred in the first few months of 1985, the statements surrounding it were rather extreme. It was stated that nobody would ever share the same genetic profile, even if you took all the people that were going to be born and all those that had been born, lived and died. This is unfortunately hopelessly incorrect, based on a misunderstanding of statistics and the low esteem in which data analysis is generally held.

Such gung-ho statements of impossibility annoyed many that had a better grasp on the statistics associated with statements of individuality. Put simply, if you can put a calculated value for the probability of an event, no matter how small that probability is, it cannot be said that it will not happen. You have after all put a value on it, so there is a finite possibility of it happening. So saying that no two people were ever going to have the same DNA profile was a statement without substance, one that could not be proven. Even given this caveat, DNA profiling was a powerful technique and one that was going to help many people come to terms with the grief and anger of rape and bereavement. But before the killer of Lynda Mann was brought to justice another vicious attack was to take place.

Nearly three years after the attack on Lynda Mann, the daughter of Barbara and Robin Ashworth, Dawn Ashworth, was to receive the same treatment at the hands of the same attacker. At fifteen Dawn Ashworth was the same age as Lynda had been when she died, and lived in a village only a short distance from Narborough. It was the very last day of July when Dawn went out in the evening to visit her friends. Dawn walked along a footpath from Enderby to Narborough, also passing close by Carlton Hayes hospital. Enderby and Narborough are two villages that are very close together, but at the time quite separate from each other. The walk takes less than twenty minutes along the country lanes that link the two villages. Narborough has a railway station that also serves Enderby and a nature reserve of reed beds and woodland on the river Soar.

In the household of Barbara and Robin there was rising

tension, until Dawn was definitely late getting home. They went out to look for her: at that moment, calling the police might have seemed a bit hasty. Dawn did actually prove to be missing and so by the morning of 1 August a search was started, but the possibility that she had simply run off could not be entirely discounted. That she had disappeared on her own, for whatever reason, was probably the best thing for Dawn's parents to think had happened at the time. The alternatives were too frightening to consider. What had happened to Lynda Mann not so very long ago in the same quiet area of the country can't have been very far from their minds.

Late that evening on 1 August, the Ashworths received an anonymous phone call. It didn't reassure them: it increased the tension and fear they felt because it was silent. There were no words and no breathing; it was impossible to know whether this was an apologetic runaway Dawn, trying to make contact, or a malicious call from a killer or kidnapper. Or perhaps it was quite simply a hoax perpetrated by an unpleasant individual. It happened again, twice, the next day. On this same day, events were going to take a turn for the worse.

At about noon on 2 August the police discovered a partially hidden body under a pile of foliage just off the lane Dawn had been seen entering. The body was partially clothed and having been left for two warm days in July there was some considerable insect damage to the body as well as damage ascribed to her having been dragged to the place of hiding. It was Dawn's father who made the official identification after which, in the early evening, the post-mortem started. Dawn had been about 165cm in height. There were many severe bruises and abrasions, most of which appeared to have been inflicted after death when the assailant attempted to hide the body. Some of these had been inflicted before death, as had the injuries to her vagina and anus during sexual intercourse involving forceful penetration. The pathologist noted that since all the damage to her vagina had been inflicted during the attack, until then she had been a virgin. Once again the cause of death was strangulation and it was possible that the sexual attack occurred at the point of death.

Very soon afterwards a local businessman offered a reward of £15,000 for information leading to the capture of the killer. Police naturally assumed that it was a single killer. It was about four weeks after the murder that Dawn was buried in Enderby. Quite soon after the body of Dawn was found an arrest was made of a kitchen porter from Carlton Hayes Hospital. Even though his PGM test should have ruled him out as the killer of Lynda Mann, the police persisted in trying to piece together a case while the porter kept denying that he had anything to do with it. He had however confessed under pressure to being the killer of Dawn Ashworth.

A step forward in clarifying the situation was presented when the police, to see if his new DNA analysis technique could be used to answer a fundamental question, approached Alec Jefferys. The question was whether the porter had killed Lynda Mann. They had his confession regarding Dawn Ashworth, so at that point the police were not interested in testing the porter against the semen recovered from the body of Dawn Ashworth: his confession was seen as enough. But they wanted to link him to the death of Lynda.

The first stage in the process was to take a blood sample from the porter that could then be compared with the DNA from the semen found on Lynda Mann. The DNA profiles of the blood and semen would be identical if they came from the same individual, but they were not.

This was a problem for the police because it quite clearly meant that the porter was not the rapist of Lynda Mann. The reason that this was a major problem for the police was because they had always assumed that it was the same man who had carried out the two attacks. If they had stopped at that point it would have to be a case of a terrible coincidence that two victims had two different attackers. The only thing to do was test material from Dawn Ashworth, in this case semen, to find out definitively if the porter was the culprit.

What the second round of DNA analysis revealed was that the two girls had indeed been raped and killed by a single attacker and it was not the porter who was being held under arrest. The porter was released entirely on the basis of DNA evidence. Not only was this the very first time such a thing had happened but also it underlined one of the greatest strengths of DNA profiling. It is possible to say categorically that someone is not responsible for a crime, regardless of any admission.

In the meantime life did not return to normal in the two villages and the police once again started a large-scale enquiry. This took a turn for the better when it was decided that every male old enough to have murdered Lynda and Dawn should be asked to voluntarily donate a blood sample for testing to rule them out of the enquiry. The announcement was made on 2 January 1987. The police wanted every male between seventeen and thirty-four to give a sample. This was to include, besides residents of Enderby and Narborough, residents of another nearby village, Littlethorpe, as well as men who had been in or around these places at the time the offences were committed. The logistics of organizing such a huge sweep of individuals was not lost on the testing teams. It became a huge logistics job of tracking everyone down, and trying to persuade him or her that giving blood was a public-spirited thing to do.

Since it was assumed, probably correctly, that the perpetrator was unlikely to volunteer to give a sample, when the original take-up from the first round was seen to be about 90%, the remaining 10% were contacted directly by police officers. This assumption by the criminal that DNA profiling was a very powerful technique may have been correct, but since this was a very new method, never before used, it was always possible that an individual may think they could bluff their way out of any incriminating result.

To protect against the possibility of fraud, a set of rules were instituted which ultimately paid off, but in the short term went against the investigation. It was not just a case of taking blood for analysis, it was also making sure that there was some record of who had given blood and whether there were any suspicious circumstances regarding an individual at the time the crimes were committed. This was done by a person taking something with a photograph on, like a passport or self-employment card (remembering that there were no photo card driving licences at this time). Failing that, the interviewing police officer would take an instant photograph. All this took time and cost a great deal of money and in the end proved ineffectual. The processing of samples also took time and cost a great deal of money, in fact it was realized that because of the very great skill needed to process DNA samples, it would be better if a preliminary screen of samples was made to eliminate as many individuals as possible. This would then only leave the a few to be screened using DNA. Since it was known that the attacker was PGM1+ secretor, if the sample was not of that designation the sample was disregarded. On the other hand, if it was then the analysis was pursued using a DNA profile.

The PGM testing was carried out at the Home Office Laboratory of the Forensic Science Service in Huntingdon, Cambridgeshire, while DNA analysis was carried out at the Home Office Laboratory of the Forensic Science Service in Aldermaston, Berkshire. The reason that it was done there rather than at the laboratory of Alec Jefferys is a simple one. His was a research laboratory and even though there had been a preliminary screen, the number of samples would have simply swamped his facilities, so a laboratory was used which was geared up to handling large numbers of samples. This inevitably meant training large numbers of staff to undertake the skilled work of DNA extraction and processing the DNA to produce a final result. This training lasted over a year, from June 1987 to July 1988. By the end of the training period 20 biologists had been trained in the practicalities of producing a profile and the underlying theoretical aspects of interpreting the results. These forensic scientists were quite likely to have to defend their conclusions in court, so they would have to be completely familiar with the technology that they were working with. At this time the forensic scientist carrying out the analysis would also write the report and if necessary appear in court to explain the results.

Fifteen years later it was quite common for analysis to be carried out in one Forensic Science Service laboratory and then the results sent to another laboratory for interpretation and the writing of the report. It is this individual, the writer of the report, that appears in court.

One of the individuals who received a letter from the police requesting a voluntary donation of blood was Colin Pitchfork. He did not want to give a sample, but knew that it would be necessary to show his innocence, so he thought of a method of circumventing the system. He would persuade a friend to give blood on his behalf. This might or might not be easy, what would be more difficult would be to circumvent the identification that the photographs would assure. It was said that if you did not have adequate photographic identification the police would take your photograph and then show it to your neighbours to make sure it was you. Colin Pitchfork had a passport, so if he could persuade someone to give blood on his behalf he would replace the photograph in his passport with one of the person who was actually giving blood.

Colin managed to persuade a colleague where he worked, Ian Kelly, to give blood on his behalf. Colin then proceeded to doctor his passport and replace his photograph with a photograph of Ian Kelly.

By half way through the year nearly 4,000 men had been tested and about 50% had been eliminated on the original screen. There had been a remarkable response to the voluntary requests for blood samples, but there was still a large pool of likely suspects who had to be contacted, persuaded and sampled before the story would be complete. Still there were no matching DNA profiles. In the end 4,582 samples had been taken and analysed with no result. It was at this point that something remarkable took place. While having a drink with a colleague, the manageress of the bakery where Ian Kelly worked, Kelly let slip that he had taken the blood test on behalf of Colin Pitchfork.

That casual remark by Ian Kelly was considered for several weeks by the bakery manageress where Ian worked until she made a phone call which set in train a series of actions which was to prove most fruitful for the police and the pursuit of justice.

The suggestion was made in the call that Colin Pitchfork had not given blood but persuaded a friend to give blood in his place. The first thing for the police to do was check the signature of the Colin Pitchfork who gave blood with the signature that had been given by Colin during the house-to-house inquiries of the Lynda Mann investigation. They were different, and alarm bells sounded. As they had the name of Ian Kelly it was him that they interviewed first, and arrested for perverting the course of justice. For this he received an eighteen-month custodial sentence, which was suspended for two years so that as long as he did not commit another criminal offence he would not have to serve his sentence.

It was later on 19 September that Colin Pitchfork was arrested, the same day as Ian Kelly was handed his own sentence. During his interview it transpired that he was a habitual 'flasher' to women, and already known to the police in this capacity. It also came out that he had indecently assaulted two other girls and attempted to abduct a third, for which he was also charged with kidnapping. When the unequivocal DNA of Colin Pitchfork was analysed it did indeed match the DNA profile generated from the semen found on both Lynda Mann and Dawn Ashworth. This data, along with his confession, resulted in his appearance in court.

When faced with the charges in court Colin Pitchfork pleaded guilty to the two murders, to the two indecent assaults and to conspiracy to pervert the course of justice by arranging another donor to give blood in his name. Strangely, he pleaded not guilty to the charge of kidnap. He received a life sentence for each of the murders, ten years for each of the rapes, three years each for the sexual assaults and three years for the conspiracy. The Lord Chief Justice later set a tariff of 30 years, so Colin

Pitchfork will not be out of prison until 2017, at the earliest.

There are some, however, who can never be released: the relatives of the victims. For them the sentence is forever. When a parent loses a child through illness or accident it is possible to take solace from the fact that it is out of human hands. When one human being exercises brutal control over another, when one more powerful subjugates another when they should be protecting them, this seems unimaginable. But during those years when Colin Pitchfork was about that is precisely what happened.

The twist in this particular tale is that although it was mass DNA profiling that brought Colin to court and proved him guilty, this was only because he had not given a sample. He had not been caught because he gave a sample, which was then linked to the two victims, but because a colleague let slip that he had given a sample on Colin's behalf, and this fact was related to a third party who had luckily realized the significance of the admission.

As a mass screening technique it was a brave first attempt, although the complexity of interpreting the results, even understanding the basic science employed, would be difficult for a jury of non-scientists. As time went on the systems used were progressively easier to understand and the arguments in court became less common, but for the next few years controversy was going to follow DNA profiling around, both here and in the USA.

Moving forward several years, to April 2004, the latest use of DNA profiling has been used to track down a murderer. While it is well known that most murders are committed domestically and the perpetrator and the victim are known to each other, sometimes the killer had never seen the victim and the death is a result of a single, thoughtless and malicious act.

It was in March 2003 when Craig Harman was crossing the M3 motorway in southern England on foot, across a bridge over this motorway. The bridge was near Camberley, not far from where Harman was living. At the time Harman was 19 years old and was living with his girlfriend in Frimley, a small town in Surrey, England. By his own admission he had been drinking heavily and was drunk. He had been out drinking with a friend; after starting their walk home they picked up a couple of bricks from a garden with the intention of using them to disrupt the traffic on the motorway, by hurling the bricks on to the road, from the bridge. When Harman threw his brick from the bridge on to the M3 motorway it hit the windscreen of a lorry being driving by Michael Little. Vehicle windscreens are generally very tough things, but if you consider the force with which a brick hits a pane of glass at 60 miles an hour (95kmh) then even laminated glass stands little chance of holding firm. The brick went through the windscreen of the lorry and hit Michael Little in the chest.

Even though severely injured, in an act of incredible responsibility, which was praised in court, he managed to steer his vehicle on to the hard shoulder of the motorway and bring it to a stop before he died of heart failure. As soon as the brick hit the windscreen, the charge against the perpetrator was criminal damage. As it passed through the windscreen and killed Michael Little, the charge changed to murder. Since Michael had managed to park his lorry before his death, it looked not so much an accident, more an apparent breakdown. It was then three and a half hours later that a policemen in a patrol car noticed his body slumped over the steering wheel of the lorry, and, reportedly, the driver's foot still on the brake.

Earlier that evening, Harman had cut his hand whilst

trying to break into a car and had left blood on the brick that he later picked up and threw on to the motorway. At that stage, Harman had never been in trouble with the police so that when the unknown DNA profile was reconstructed from the bloodstain on the brick it produced no results, when compared to the UK national database. The question for the police was where they went now, with an apparently unprovoked attack and no suspects. A reward was offered of £25,000 for information that resulted in a conviction. This produced no new information, however. What did make a difference was when the Home Office Forensic Science Service contacted Detective Chief Inspector Graham Hill. The suggestion was that by using a large number of short tandem repeats (STRs) it might be possible to track down a family member. Now, this is a very dangerous business to get into, and although this worked well in this particular case, great caution should be exercised when using this technique, as we will see.

When confronted with the evidence that the STRs provided, it still took three hours of interrogation to get Harman to admit to the lesser charge of manslaughter. He had been picked up and a DNA sample taken which exactly matched the DNA profile found on the brick that had been unthinkingly hurled through the window of the lorry driven by Michael Little.

The STRs pointed to Harman because a close relative of Craig Harman had already been charged with an arrestable offence; consequently, the DNA profile was on the national DNA database. This is where it becomes increasingly tricky to be precise and sort the wheat from the chaff. We are not identifying the perpetrator directly from his DNA profile. What we are doing is finding a DNA profile that matches in part the profile of the bloodstain found on the brick. It is axiomatic that this is not itself enough to produce a partial profile, which could be used to drag an individual in, off the streets to try to force a confession. What is required is additional information, which is presumably what took place in this case. Once the profile from the brick was matched to a close relative, it was possible to tic in all those individuals that could have perpetrated the crime. Once arrested, it was possible to demand a sample from Craig Harman for DNA profiling which could then be directly compared to the profile taken from the blood on the brick. When using new technology great care must be exercised. It is not possible to assume guilt without additional information and evidence.

Such a situation had been found where there were either deliberate or accidental mistakes made which threw into question the validity of some – several – convictions based on DNA evidence. The events that very nearly undermined public confidence in DNA profiling in the USA took place in Texas.

Early in 2004, it had been discovered that the Houston Police Department Crime Laboratory had been using some very dubious scientific practices in the pursuit of criminal cases involving DNA. It meant that it was made necessary to sift through all the DNA cases since 1992 for signs of errors or misconduct. The major problem for the officials in Texas is that this state hands out more death penalties than any other in the USA. It has been such a disaster for Texas crime laboratories that it has been suggested that extremely strict regulation for labs carrying out this work should be enacted as legislation. As a result, it is also progressively more common for non-scientific and scientific groups to suggest that Texas is not going to be an isolated case. The misuse of DNA profiling could be far more widespread, they say.

Without detailed analysis it may never be known whether these mistakes were the result of ineptitude, stupidity, laziness, or even a belief by the technicians carrying out the analysis that they were helping out society by finding a match that does not exist. While the original point was raised in Harris County, Fort Worth police department announced a review of all cases where DNA was pivotal in proceedings. This resulted in the resignation of an FBI technician who was being investigated for not following correct DNA analysis procedures. Such was the consternation at the scale of the problem that the Mayor of Houston called for a moratorium on all executions until the evidence had been reviewed.

Many of these cases under review involved the death penalty, but by their very nature, they all carried huge tariffs. It was these cases which shook public confidence in DNA testing, a confidence founded on the belief that DNA was an unbiased and reliable form of evidence. DNA evidence is indeed unbiased and reliable, but only when carried out correctly; no technique is so good that it cannot be ruined by inappropriate actions on the part of the technicians carrying out the procedure. In the USA, at the time that the Houston police department was having all its problems, accreditation of forensic laboratories was voluntary in all states with the exception of New York and Oklahoma, where it is mandatory. In most cases, the laboratories carrying out these forensic analyses of DNA did not bother to accredit their laboratories, leaving them legitimately open to severe criticism.

One such case involved a 25-year sentence. Josiah Sutton received this sentence based on eyewitness testimony and DNA. While we cannot underestimate the damage that crimes such as rape cause for the victim and their families, we always need to be sure that the right person is brought to account for their actions.

During one night in October 1998, a woman was abducted at gunpoint from the car park of her own apartment block and driven away in her own car. At an isolated spot the car was stopped and both the men systematically raped the woman. It was nearly a week later, about five days in fact, that the victim was passing a group of people and claimed to have seen the two people responsible for her degrading assault. One of these two individuals was Josiah Sutton, who was 16 years old. The police were seemingly looking for an individual of about 1.75 metres in height and of about 61kg in weight. Josiah Sutton was nearly 2 metres tall and nearer to 90kg in weight.

Once arrested, both Josiah Sutton and his friend were interrogated and denied all charges, claiming to be completely innocent. This is nothing unusual of course – most suspects claim to be innocent. It seems that it was only when they had been put in jail that they found out that the charges were both kidnap and rape. The mother of Josiah Sutton was convinced that this was a case of mistaken identity.

Such was the conviction of Josiah Sutton of his innocence that he was convinced that a DNA sample would demonstrate his innocence. A recurring point here is that the burden of proof seems to be moving away from innocent until proven guilt towards a reverse of this. Regardless of the description given to the police not really matching that of Josiah Sutton, the police thought that they had their man. Sutton volunteered a sample for DNA profile believing that the result would produce complete exclusion and exonerate him from the charges. A sample had also been taken from the friend of Sutton who had been arrested with him. The samples from the two men were to be compared with samples of semen recovered from the victim and the back seat of her car where the rape had taken place.

It took two months for the results to be produced and there was a surprise for Josiah Sutton. His friend was ruled out as one of the rapists, but he was not. Remember that exclusion is an absolute statement that DNA does not match – and therefore individuals do not match. An inclusion however requires some sort of statement as to the probability of the match having occurred by chance. The laboratory report was described as definitive; Sutton's friend was exonerated and released, but Sutton himself was linked by a DNA profile to the rape.

At this point it is worth thinking about how the profile was produced, or, more precisely, interpreted. Semen samples from multiple rapes, which are then used to produce a DNA profile, are very difficult to interpret. The problem is a simple one to state, but the solution is extremely hard to find. If you have the profiles of two individuals mixed together, how can you possibly know which point on the DNA profile belongs to which individual? The short answer is that you cannot. The court has to decide on the balance of probabilities whether the forensic scientist had made a true interpretation of the results or not. Even worse, they have to decide whether a process has been carried out correctly, or not. This latter point is probably more that can be expected for a layman jury to understand. So after two months of Sutton hearing that there was DNA evidence against him, he was still confident of his innocence being proved, but this was quite to the contrary of the forensic report that was presented to both him and the defence lawyer representing him. It was apparently stated that the lab had produced a definitive result, with a probability of finding the same profile within an unrelated population of about 700,000 to 1. This is an easy number to produce, but a very much more difficult one to defend. If a jury takes the date to be true at face value, the outcome could be devastating to the individual.

The jury took less than two hours to decide that Josiah Sutton was guilty, with the judge consequently giving Sutton a sentence of 25 years. In a state of disbelief, Sutton started his sentence, but he did not just sit down and take it. He knew he was innocent and so set out to find a chink in the armour of the prosecution. The way he did that was to study that which had convicted him – DNA.

Interestingly, it was media that turned the corner for Sutton. His lawyers decided to speak to two reporters who worked for a television station based in Houston called KHOU-TV. The two reporters were Anna Werner and David Raziq. What they managed to do was ferret out the reports from the court and forensic laboratories that carried out the original DNA testing. They sent all of the documents to a wide range of experts, including a professor of criminology at the University of California, who reviewed DNA evidence from all over the country. The result was a consensus of opinion, signalling a difficult conundrum. By common consent it was agreed that not only was the DNA profile of Sutton incorrect, but it implied that the entire work of the lab was flawed. With a suggestion of either incompetence or complicity with the authorities, the first criticism was in the way that statistical interpretation had been made of the results. It got worse. Without even having to carry out another DNA analysis it was becoming clear that the whole procedure, as stated by the prosecution, was horribly flawed. So flawed, in fact, that from the original profile that had been produced, it was suggested that not only was it a case of misinterpretation of the statistics, it was a case of suggesting an inclusion when it should have been clearly seen for what it was – an exclusion.

Analysis had been carried out on the semen stain from the back seat of the car. The crime lab had made an incorrect profile. They were wrong, so wrong that the police were persuaded to look again at the DNA and conduct retests of the samples. The retest was supervised for the defence to make sure that the original mistakes were not replicated. The general feeling was that when forensic science labs are located in police departments the result is usually a partisan outlook that can all too often end up with forensic scientists being asked to produce a result that will match police expectations. The police and forensic scientists are just too close together. Also, as Peter Neufeld, one of the defence team at the O.J. Simpson said in an interview, forensic science is an oxymoron.

Independent scientists who assessed the laboratory said that incompetence had been covered over by bias. The final report concluded that from inadequate training to deliberate tampering with evidence, the lab was hopeless, and so it was closed down. Even though Sutton was released after four years, the judge in the case refused to absolve him from the guilty verdict. Even though the DNA evidence was not only flawed but also indicated that the DNA was not even there on the back seat of the car, the judge, pending ongoing investigation, withheld Sutton's absolution.

Although mass DNA screening is occasionally carried out, it is still relatively unusual. A case where it has been done involves a small community in Australia. Mass screening is really only a feasible option in these small communities, such as the previous case of Narborough and the following case of Wee Waa in Australia. It was on

New Year's Eve in 1998 that a 91-year-old was attacked, assaulted and raped in a small town northwest of Sydney in New South Wales, Australia. Wee Waa is a small town and until that day had a very low and parochial crime rate. It was easily the most vicious and easily the worst crime ever having been perpetrated there.

There were lengthy investigations carried out by the police, but no result was produced. There were no arrests, not even suspects. The thought was that it might have been an itinerant worker, or passing traveller. Still, there were no leads that could result in an arrest, or better still a conviction.

Help was eventually requested from the local community to help find the culprit. A reward of A$50,000 was put up to help in the process, but what was more important was that if the criminal was a local man, they could all be tested for DNA. There was a sample against which comparison could be made from the crime – which would lead to a conviction. No culprit has been found as yet.

4 THE ROYAL FAMILIES OF EUROPE

One of the most interesting things, for geneticists, about the Royal Families of Europe is that they have very well documented family trees. This is of considerable importance when we come to look at the process of centuries of what is, in effect, inbreeding. Odd aberrations and telltale signs have given us a surprising clue to the method in which genetic disease is inherited.

Admittedly it is possible to research many families in the UK because of the extensive records that have been kept for hundreds of years, but it is not easy or accurate. Put another way, we have a good idea who has been responsible for most of the people that have been in our royal families, wanted or not, for several centuries. This detailed catalogue of the European royal families tells us much about the nature of hereditary diseases, some of which have appeared and disappeared, but are all well documented. This is in contrast to the records of many peoples' families, for whom the trail of ancestors normally runs out before the halt caused by the widespread destruction of records during the English Civil War (1640-1648). Details of the royal families were written down in so many different places, from genealogies to history books, that it would be extremely difficult to lose a member.

George III was born on 4 June 1738 and died 29 January 1820. He was the King of Great Britain and is best remembered for having presided over the loss of what were the American Colonies, and now the USA. However, there is another question that we can look at, and potentially answer in retrospect. This is the question of his illness, 'the madness of King George'. Any ability to diagnose a condition retrospectively that we have is based upon the incredible ability of contemporary commentators in describing George's symptoms. About 1765 he seems to have suffered a bout of dementia, recurring in 1788. This later illness was so severe that a Bill was passed to give the regency to his son. However, George III recovered and went on, until the situation became progressively worse, with blindness in 1809 and by 1811 complete dementia resulting in his son, later George IV, taking over the regency.

Many families have stories about 'odd' family members. Perhaps they died young for unknown reasons. This is often referred to as the family curse, turning up from time to time in different generations. I have seen this myself, in families where there is a history of stillborn babies, or babies that only survived for a few hours or days. Looking at the family members directly involved, such as the parents, brothers, sisters and if possible grandparents, it was often possible to discover a genetic problem. It would then transpire that the older generations had known that there was a problem, with relatives known to have died in childhood, or with a poor history of pregnancy, but had chosen to remain silent because they thought it reflected badly upon the family. Just like you cannot choose your family, you cannot choose your genes – just yet.

In the case of George III, the details of his condition were both well documented and well observed, allowing us to make a good guess that what he suffered from was a form of porphyria. In fact it is even possible to suggest something even more specific: it is thought that he actually had intermittent acute porphyria.

All of the different forms of porphyria are associated with the production of haemoglobin from a most important procursor porphyrin. Tho importanco of this molocule can be seen from the fact that in broad outline if porphyrin combines with magnesium it forms the green pigment chlorophyll, which plants use to capture sunlight and generate sugars and other carbohydrates, which are so important in sustaining life. If the porphyrin combines with iron it forms haem. Once the haem combines with a variety of proteins it forms a group of molecules that are very important for animals and the most important of these is haemoglobin, the red pigment of blood that conveys oxygen to the tissues of the body, and carries poisonous carbon dioxide away. It is this importance that results in a simple problem having such a profound result.

The various forms of porphyria turn up at a rate of about 1 in 50,000 in north European populations, but peculiarly in white South Africans it rises to almost 1 in 1,000. This disparity in numbers is probably due to a mutation having occurred and been passed amongst a relatively small founder population.

It is thought that intermittent acute porphyria is generally inherited in a dominant manner (that it is passed through the generations on a dominant gene) but the story is rather more complicated because it does not seem to be manifest in all people at the same rate. It is a very sensitive disorder; sometimes it seems to remain dormant for generations, but can be triggered by a variety of chemical and hormonal assaults. These include several prescription drugs, alcohol, changes in the hormonal balance, as in the menstrual cycle, or the use of hormonal contraceptive pills.

This is probably why women have a five times more frequent attack rate than men. Symptoms tend to be intermittent, involving neurological disorders, the digestive system and skin problems. The attacks tend to start with severe abdominal pain and because they are often late in onset, in people in their 30s for example, it is not unknown for individuals to undergo unnecessary surgery because of an initially incorrect diagnosis. In the case of George III he seems to have been struck by one of the other manifestations (psychiatric problems) that can persist between attacks. This is why patients are sometimes diagnosed with a psychiatric disorder, hence the 'madness of King George'. This sort of disorder is an intermittent problem, but some genetic conditions have a far more readily predictable way of being passed down through the generations. Of course, some problems are not always inherited from previous generations: they turn up as spontaneous mutations.

A case of this is Duchenne muscular dystrophy. Although often inherited maternally, there are approximately 40% of cases that are new mutations. The defective gene is found on the X chromosome and is recessive. This means that since girls have two X chromosomes, and boys only one, the other sex chromosome being a much smaller and virtually inert Y, it is normally only boys that are affected.

A very similar situation is found in haemophilia, where the mode of inheritance is also sex-linked and therefore not normally expressed in girls. Muscular dystrophy, where onset is early and invariably lethal, tends not to allow a female carrier and affected male to produce children, which might result in an affected female child. This situation is more likely to occur in haemophilia, but in haemophilia the complications of puberty in girls render the condition lethal for obvious reasons. The condition is generally, therefore, transmitted *via* the maternal line, but not generally expressed in it. There is another well-documented case of royal transmission and we even have an indication of where the mutation took place. It involves almost all of the royal families of Europe, started during the nineteenth century, and involves the transmission of haemophilia.

Queen Victoria was born in 1819, being christened Alexandrina Victoria. She reigned from 1837 until her death in 1901. In the interim period she gave her name to an entire era and people, the Victorians, and ruled over the British Empire at the height of its power and influence. Having lived through the bulk of the nineteenth century, Queen Victoria was arguably the most well known person in the world. She was born the daughter of Victoria Mary Louisa and Edward Augustus, who was the Duke of Kent and fourth son of George III and youngest brother of William IV, who was the monarch at the time. When William IV died he had no legitimate heirs and so Victoria became queen at the age of 18. Unknown, and taking the throne at a time where the royal family was held in low esteem by the public, without the expert and sensitive guidance of the ageing Whig prime minister William Lamb, 2nd Viscount Melbourne, it is unlikely that Victoria could have raised the royal profile to the level she did. In fact, she took it to a level greater than any monarch since Elizabeth I.

Such was the power and influence of Victoria, that after her marriage to Albert, Prince of Saxe-Coburg-Gotha in 1840, the production of nine children resulted in a series of marriages across Europe. This is important because until Queen Victoria there does not seem to have been any history of haemophilia in the Royal Family. The assumption, therefore, is that Queen Victoria was the product of a mutation on an X chromosome from one of her two parents, which resulted in haemophilia in the male descendants and carrier status amongst the female descendants of Queen Victoria. The first of the nine children was also called Victoria and became Empress of Germany. The second was the future Edward VII. From a total of 9 children she had 40 grandchildren.

So, running chronologically these nine children were –

1840 Victoria

1841 The Prince of Wales, later Edward VII

1843 Princess Alice, later Grand Duchess of Hesse; we need to keep a weather eye on this one as she is most definitely a carrier of haemophilia

1844 Prince Alfred, Duke of Edinburgh and Duke of Saxe-Coburg-Gotha

1846 Princess Helena (Princess Christian)

1848 Princess Louise (Duchess of Argyle)

1850 Prince Arthur (Duke of Connaught)

1853 Prince Leopold (Duke of Albany), who died in 1884, a haemophiliac

1857 Princess Beatrice (Princess Henry of Battenberg), another female carrier of haemophilia

So what happened after this? The first grandchild of Victoria was born in 1859 and the first great grandchild was born in 1879, more than 20 years before the death of Queen Victoria. Now, of these 9 children there was one affected haemophiliac boy, Leopold, Duke of Albany, and two carrier girls. Unusually for a haemophiliac at that time, Leopold married and had two children, one girl and one boy. The result of this is that any boys which are born gain their X chromosome from their mother and it is the X chromosome which carries the haemophilia gene, but the boys only get a Y chromosome from their father, so the affected haemophilia gene is not passed from father to son. It is however passed on to all the daughters; so all girls of a haemophiliac father will be carriers, carriers being unaffected but able to pass on the defective gene to future generations. At the next generation, of the carrier girls, 50% of the female children will be carriers and 50% of the boy children of these carrier girls will be affected, but not carriers, bearing in mind that you cannot have a male carrier of haemophilia – only completely healthy or affected.

Now this is of significance for the next stage in the story of the European royal families, because there were several descendants who were either carriers or affected individuals. Affected individuals included Fredrick of Hesse, Lord Leopold of Battenberg and Prince Maurice of Battenberg. There were also in the next generation Waldemar of Prussia, Henry of Prussia, Tsaravitch Alexis of Russia, Rupert, Viscount Trematon, and Alfonso of Spain. There is one other possibly affected child, the brother of Viscount Trematon, Gonzolo of Spain, who died in childhood.

At this point we can concentrate on the Russian royal family. Queen Victoria had a daughter Alice (1843-1878) who was a carrier. She in turn passed on the haemophilia to Alexandra (Alix) Feodorovna (1872-1918). She was a German princess who married Tsar Nicholas II of Russia in 1894, making her Empress of Russia. The wedding was a bit of a disaster. It took place at the deathbed of his father and because officials had not paid attention to the tradition of distributing 'bounties' there was a stampede with a reported 3000 individuals crushed to death. Nicholas II also managed to drop the Imperial Chain.

Alexandra was deeply pious and superstitious. As a consequence of this she came under the influence of

Rasputin, as he seemed able to ease the bleeding of the crown prince Alexis. Rasputin was a self-styled mystic and holy man, eventually killed in 1916 by a group of aristocrats. Alexandra became involved in politics during the First World War and it was openly said in Russia that she was passing information to the Kaiser, being originally German. All this and the rather hard political line of the Emperor, along with poor military results, ended with Nicholas being given an abdication document to sign, which he did with composure. The immediate result of the abdication was his internment for many months in Tobol'sk, in the southern Ural mountains, with his wife, son and four daughters. This situation could not be continued because there were rumours that the Czechoslovakian forces were advancing on the Urals. Also in the Ural mountains was a much larger town than Tobol'sk, called Yekaterinburg. It was here that the Romanovs were finally killed. Yekaterinburg was founded in 1721 by Peter I and was named in honour of his wife Catherine I. Interestingly this association with the Russian royal family was so frowned upon during the communist era of the USSR that between 1924 and 1991 the city was renamed Sverdlovsk, after the Soviet leader Yakov Sverdlov.

In 1918 the communists of Yekaterinburg in the Ural mountains, held a secret meeting at which they decided that the Tzar and his family should be killed, a message to the effect of which was sent to the Yekaterinburg commissar, by the name of Yourkovsky. The commissar insisted that the warrant was signed. This was duly done and on 16 July, Yourkovsky roused the prisoners and after reading the charge he personally shot the Tzar. The remaining guards then shot the rest of the royal household.

Now we have to fast-forward the story by 70 years. The individuals that had been shot had been dumped down a mineshaft and later removed and buried. After searching through records in Russia, a shallow grave was found in the Urals, not far from Yekaterinburg. At this point the question arose as to whether these individuals were really the Romanovs. With the assistance of Russian and UK Forensic Science Service personnel is was possible to set to and address this question. But how would it be possible to answer a question like that? The corpses were decayed, in a state of virtual disintegration. It was apparent that they had been murdered, too. The question was how is it possible to state with certainty that these bodies were related, let alone the bodies of any given family? The answer was to try and find some DNA in the bones that were left.

With co-operation between the forensic scientists of both the United Kingdom and the political authorities of the Urals, it was suddenly possible for a story to unfold about the last surviving members of the royal family of Russia. The analysis was carried out by two teams, led by Peter Gill in the UK and Pavel Ivanov from the Engelhardt Institute of Molecular Biology in Moscow. But there is more: the possible survival of Princess Anastasia.

After extraction and analysis of DNA from the bones of the individuals, using the STR process, it was possible to say with some certainty that what had been found was a family group. At this early stage it was not possible to say which family. Perhaps it was not so surprising, because of the social nature of the way in which Russian royal families would have been buried, that is, together. Unfortunately, the family group seemed to be rather extended. What had been found in the grave was a family group of a father, mother and three siblings. Besides this family there were remains of four other, unrelated, adults; nine bodies in all. These other bodies were reputedly three servants and the Romanov doctor. Having found that this group was a family group, the question was *which* family group? How was it going to be possible to decide who was who in that family group? Not only that, but there were several unrelated individuals that had also been killed at the same time, or certainly close to it and dumped in the same grave in the same ignominious manner.

Initially the method used for this analysis was short tandem repeats on DNA extracted from the long bones. STR analysis was the method of choice because it would not be expected that there would be sufficient DNA, of long enough sequences, to use any of the older DNA analysis techniques. The possibility of extracting DNA from remains which are the best part of 70 years old is very hit and miss, even though it is possible to go back in time many thousand of years and still extract useable DNA. It does depend to a large degree on the conditions where the bodies are buried. As a general rule, as bodies decompose, DNA breaks down in the soft tissues first. These are of course the skin and internal organs. After this, the DNA in bones will start to break down. DNA is actually a very robust molecule: it takes quite a long time for the molecule to be completely destroyed, but it will be broken up into smaller and smaller pieces over time.

DNA in bones, and here it is nuclear DNA that I refer to, is not generally found in the solid material, but is found in the centre of the long bones of the arms and legs, specifically the thighs. This is because these are the bones that contain the stem cells, which produce all the different types of blood cells – that is, the bone marrow. Since bone marrow is encapsulated within these long bones it is very well protected in the short term. In the long term, survival of usable DNA depends upon several factors, such as the acidity of the soil, which will affect decomposition rates and therefore the time it takes to break up DNA. Another factor of major importance is the free water content of the ground. Now, I say free water because while it is obvious that in, say, deserts the amount of water available for bacterial decomposition (which always needs water) is very limited, there is another environment which has *plenty* of water, but it is water that is unavailable for use. This is in very cold areas with permafrost, such as in the Arctic.

When there is little water because of heat, like deserts, it is still possible that carrion eaters will carry off parts, or all of a carcass, in fact. It is not unknown for carrion eaters, such as hyenas, to dig up shallow graves and eat any bodies that they find. The reason scavengers can identify such sites is by smell, even though in hot deserts putrefaction eventually turns to mummification. In the meantime rotting will start internally from the natural bacteria of decay found in the gut, not to mention the natural enzymes in the body, which are normally used to digest food, but will start to digest the internal organs until the enzymes themselves are deactivated.

An alternative to the drying effect of deserts to preserve bodies is the freezing of bodies. This has two distinct effects: the first is that metabolic activity of gut bacteria slows right down. Also, even bacteria are not immune to freezing and thawing cycles, which damage their cell walls. Even if they *were* active at very low temperatures, there would be no available water, so their activity would be incredibly slow. These areas are effectively deserts with water, but the water is unusable as it is frozen solid.

In the case of areas of extremely low temperatures, such as the Urals, decay of corpses can go through periods of low and high temperatures, or more precisely the air temperature can go through a wide range of temperatures. Once a body is well below ground level, even in a shallow grave, the temperature changes very slowly season to season. It might sometimes be subject to decay, but it can be preserved very well.

STR analysis revealed a family group, but, of course, it was at this stage only non-DNA evidence that suggested that these were the Romanovs, especially since there were two members missing. One of the results of analysis is that it is possible to determine the sex of the DNA donor, so if they were the Romanovs, the two missing individuals were a male and a female. Since it is possible to determine the sex of an individual from a skeleton, as well as a broad idea of the age of the individual, it was possible to say that the family group comprised a mother, a father and three female siblings.

The method used to decide whether these were in truth the remains of the Romanovs was by complicated family analysis. This involved analysis of mitochondrial DNA (mtDNA), which is inherited exclusively through the maternal line of descent. From a brief review of the family tree of the royal families of Europe it was apparent that the Duke of Edinburgh shared a common maternal line of descent with the Tsarina. By analysis of the mtDNA it was seen that the female members of the family in the grave were indeed related to the Duke of Edinburgh, which added additional weight to the idea that these were the last of the Russian royal family. The final and convincing evidence came when the known remains of Grand Duke of Russia, Georgij Romanov, were exhumed. The importance of this was that he was the brother of Tsar Nicholas II, so having shared the same mother they would have the same mtDNA, if they were in fact brothers. They did, so people were convinced that the five bodies that had been exhumed were definitely the Romanovs.

But there were a number of questions that were raised

by these results. The first was where Prince Alexis had disappeared. This was easily answered: as a haemophiliac he was almost certainly dead from his own genetic condition. Although survival might have been possible, bleeding into joints and consequent tissue damage would be fatal if not treated and exacerbated by rough treatment. That left Princess Anastasia unaccounted for, so where had the last missing child of the Romanov family gone? It was unknown at the time, and we still do not know for sure, but a suggestion was made that Anastasia, Princess of the Russian royal family, had survived. This comprises the final element in our story.

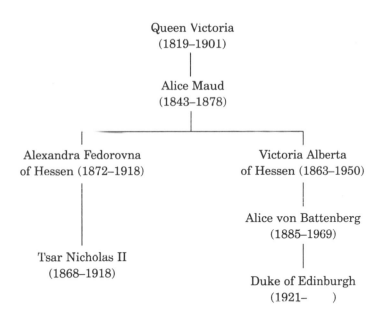

Fig. 6 The Victorian royal family tree

By looking at mitochondrial DNA (mtDNA) – which is only inherited maternally – it is possible to show that the Duke of Edinburgh should have the same mtDNA originally inherited from Queen Victoria, as Tzar Nicholas II.

Since Anastasia was the daughter of the Tzarina, and HRH Duke of Edinburgh was a direct descendant of the Tzarina on his mother's side, Anastasia would have the same mtDNA as the Duke of Edinburgh. It had been shown that there were some mutations that had taken place in the mt(DNA) over the intervening generations, but given the constraints of the technique it would be possible to check whether anyone suggesting they were Anastasia really was that person.

There was, in fact, an individual who claimed to be Anastasia. This individual was called Anna Anderson and her story started in Berlin in 1921 when she was in a mental hospital and claimed to be Anastasia. Now, unfortunately Anna Anderson died in 1984 at Charlottesville, Virginia, in America, before DNA analysis in any form was available. But the extraordinary ability of hospitals to store materials came to the aid of those who wanted to get to the bottom of the story. This meant that there were samples which, as always in health services, were clearly labelled and of known origin. The first recognized sample came from a doctor in Germany who had retained a blood sample from Anna Anderson as a souvenir. Also Anna Anderson had been taken into the Martha Jefferson Hospital for surgery in 1979, at which point a small bowel biopsy had been taken. Altogether there were five biopsy samples, of which two finally reached the laboratories that were going to carry out the analysis.

Like all such biopsies, they are taken primarily for microscopic analysis. To use them for microscopy to investigate a medical condition, they have to be embedded in either hard wax or plastic resin and then stained, to show the different cell types that are present. Before this embedding takes place the sample is 'fixed' in formaldehyde, which cross links the proteins in the tissue, rendering it relatively hard so that the internal structure of the specimen is not lost. After that the water is replaced by a liquid, which allows the permeation of the sample by the wax. The result of all this is that the cellular structure of the sample is maintained, which includes the nucleus and with the nucleus the DNA of the cell. Luckily DNA is very robust so that all this manipulation does not destroy the DNA: in fact it helps preserve it. So even after ten years in a cupboard in a pathology laboratory, it was possible to extract DNA from these samples for STR analysis. Slices only six microns (six thousandths of a millimetre) thick were taken from the blocks and it was found that the DNA was degraded to fragments of only 250bp long. The analysis of this material from the biopsies clearly demonstrated that it was not possible that Anna Anderson was Anastasia, indeed not a Romanov at all.

The question remained, however, as to who was Anna Anderson. There was a suggestion that she was actually a Pole, by the name of Franzisca Schanzkowska. But how could this be demonstrated? There was a documented great-nephew of Franzisca Schanzkowska, called Karl Maucher, related through the maternal line. So it was possible to test mtDNA and find out whether this matched between Anna and her putative family relative. This could be done using hair samples. A local historian, Susan Burkhart, was allowed access to the remaining estate of Anna, and she came across an envelope, which contained a lock of her hair. Hair shafts do not contain nuclear DNA, but they do contain mitochondrial DNA and it was this that was analysed at Pennsylvania State University.

Altogether three different laboratories analysed various samples. These were the Anthropology Institute of the University of Goettingen, the Home Office Forensic Science Laboratory in the UK and Pennsylvania

University. Much to the surprise and delight of all concerned the results all matched. The surprise was because each of the testing laboratories did not know that the other laboratories were working on the same material. What they showed was that there was now little doubt that Anna Anderson was actually Franzisca Schanzkowska, a Pole, not even a Russian. That there was a match between these two people, the aunt and nephew, does not in itself constitute proof of origin, but it does demonstrate that she was not Anastasia – that much is certain.

This is absolute: ruling out an individual as a relative, an heir, or antecedent cannot be gainsaid. It is only when an inclusion is found that arguments can start between those that claim that this proves a family relationship. It is a semantic problem, involving the word 'proves' or 'proof' that has been the cause of so much misunderstanding. For example, if you have the 'pleasure' of travelling in the UK by train to Birmingham from London, you will pass by a large and well decorated building which proudly proclaims on its roof that this is the Gun Barrel Proof House. In the modern idiom this building would now be known as the Gun Barrel Testing House. A DNA test is something that does not prove a relationship, as we know it in modern language, but *tests* the relationship – it offers a change to *proof* it. We can never be absolutely certain that a DNA sample shows a relationship, but we can proof it. The only certainty is when it rules out a possible relationship. This is when the proofing offered by DNA actually *proves* something.

An example of the proofing providing a proof was the set of results produced in 1994. They did not find universal acceptance. Supporters of the idea that Anna Anderson was really Anastasia were not going to give up so easily. As far as they were concerned they did not necessarily have to dispute the scientific findings. It was pointed out quite correctly by them that there was no documented chain of custody of the samples. It is not unknown for sample labels to get lost or for samples to be mixed up with other samples, and the hair samples were assumed to belong to Anna, but they might not have been. It is true, however, that since all the analyses matched each other, the chance of this happening among three different laboratories is remote in the extreme. What was contested was the conclusion that was drawn from the results. On balance the scientific evidence does seem to confirm the idea that Anna Anderson was not Anastasia. The detractors of the DNA evidence now seem to have quietly gone away: eventually the amassed evidence cannot be gainsaid. With no contrary evidence, it was quietly accepted.

Ten years later, in 2004, a postscript to the Romanov story was published. The criticism followed the line that the original DNA analysis was flawed. The Russian Expert Commission Abroad supported this idea. This is an expatriate group who are determined to believe that the remains were not those of the Romanovs. They were allowed to analyse DNA from Grand Duchess Elisabeth who was the sister of the Tzarina. The critical conclusion was that the remains might not even have been those of the Romanovs. The conclusion we draw from this is that since there was always going to be a residual doubt about the original studies, as there always is in DNA analysis, this may or may not add to the probity of the original analysis. This is likely to become a political question with the involvement of expatriate Russian groups and a statement from the government of President Putin that says the case is closed.

The residual doubts stem from the way the bodies were dealt with. It seems highly likely that the bodies were a family. It is certainly within the ability of DNA analysis to say that five of the individuals were related. So of the nine bodies, five were related, four were not; if they were the Romanov family there were two bodies missing. At this point we have an agreed family, and it was the mtDNA that seemed to indicate that they were the Romanovs. Strictly speaking, the mtDNA analysis showed a mismatch, but it was reported that the probability of the remains not being the Romanovs was remote.

Further analysis of other descendants and antecedents of the Romanovs seemed to show that this mismatch in mtDNA was generally acceptable, being within the range of change that would be expected over the number of generations that had passed since the earliest individual was tested. Even so the criticisms just do not seem to go away, but now they seem to have shifted towards suggestions of contamination of the ancient DNA with modern DNA. Whatever the truth, I cannot help but think that other than as a curiosity it will not change anything, certainly not for the Romanovs.

5 THOMAS JEFFERSON: A STORY OF AN INVESTIGATION THROUGH HISTORY

The suggestion that Thomas Jefferson fathered a child with one of his slaves is a controversial one. Rumours of this started in his lifetime and have not gone away in the intervening 250 or so years. This is a question that we can at least address using DNA, even if we cannot definitively answer it.

Thomas Jefferson was the third president of the United States of America. He was born at Shadwell, Virginia, on 13 April 1743, and died on 4 July 1826. He was, without doubt, tall for his time, at 6ft 2in, the best part of two metres. He was, literally, head and shoulders above most of his peers. He was regarded as rather unattractive in his youth, even gangly and undignified. This attitude towards him seems to have mellowed in his later years, reportedly because of his easy charm and ease of manner.

The question of slavery and the conjunction of slavery with important historical figures such as Thomas Jefferson may not necessarily sit comfortably in history. It is the same when questioning the political tendencies of famous artists or writers. The work of Paul de Man in critical theory is often set alongside the revelations that he was involved with the Nazis at an early stage in his career. Some literary figures of the 1920s (T. S. Eliot, Ezra Pound, Wyndham Lewis) harboured right-wing tendencies. What then should be made of the body of work, or, with a politician, the body of action, left over? It is a difficult issue to reconcile. It is important that it is put into a correct perspective.

Slavery has had a long tradition and though we would not condone it now, it was accepted in many parts of the world, until not too long ago. When dealing with the actions or the art of people in the past involved in slavery, in right-wing politics or other sorts of extremism, we can follow two broad approaches. One is to separate completely the man or woman and their body of work. This results in a structural kind of analysis, one not complicated by outside moralities or politics. Other approaches however will always bring in some element of having to reconcile what the beliefs add to or detract from the work.

Slavery has not disappeared. The slavery that we learn of in history books has been outlawed and eradicated, but the actual act of enslaving an individual is still found in the world today. Western civilizations have outlawed slavery but not entirely eradicated it from the world. Primitive hunter-gatherers would not enslave vanquished tribes, they would slay them. The society did not want or need slaves. Eventually we find change towards the removal of unnecessary toil from the vanquishing tribe by taking the vanquished and putting them to work. Once a more sedentary society was established it was necessary that the wealthy could continue their various activities, which did not involve manual labour. Once someone hits upon an idea that offers immediate material and temporal gain (money is earned and time is saved) it takes a huge amount of protestation to end it, even if it is immoral.

Throughout the eighteenth century some politicians were vehemently against slavery. William Wilberforce was definitely against the whole practice, but it was on 2 April 1792 that William Pitt, who had become Prime Minister of Great Britain in 1783 at the age of 24, made a moving speech. This was fiercely against the idea of slavery:

and I shall oppose to the utmost every proposition which in any way may tend either to prevent or postpone for an hour the total abolition of the slave trade....

It was one year after the death of William Pitt in 1806 that slavery was finally abolished in the United Kingdom. However, Great Britain was not the first European country to abolish slavery; that honour goes to Demark, which got rid of slavery in 1792. The gradual abolition of slavery proceeded with British India in 1843, Korea in 1894, China in 1906 and eventually the Arabian Peninsula in 1962. In the USA the 1862 decree of President Lincoln, against slavery, was finally written as the 13th amendment of the US Constitution in 1865.

So how did Thomas Jefferson come to be a slave owner? If we accept that morality is absolute, he was wrong, but the analysis of 'morality' often has to be temporal in order not to condemn outright entire social groups through history. He had been a member of a slave-owning family and so ownership of slaves was not seen as wrong, even though this was more that 50 years since Great Britain had outlawed slave ownership.

Thomas Jefferson lived in Shadwell, Virginia. On the death of his father, Thomas found himself as the inheritor of 1,900 acres, which left him with a sufficient income to make him a man of independent means. By the time he was 30 he had increased this to 5,000 acres. He married a wealthy widow, Martha, of only 23 years of age, in 1772. When his father-in-law died the next year in 1773 he inherited almost as much again, although this estate was heavily in debt. He was admitted to the Bar in 1767, but during his late twenties his greatest passion was the design and building of what was going to become his life-long home on the edge of Charlottesville, which he named Monticello.

He started his public life as a justice of the peace. Later he was a theoretical pursuer of independence for the North Americas and principal author of the 1776 Declaration of Independence. Here we have a huge anomaly in Jefferson. He appears to believe that what is right for the ruling class is not necessarily right for everyone, but then, this is a tenet of Western philosophy passed down straight from Plato and Socrates. Their idea of the philosopher-kings, artisans and slaves, a kind of Junior Common Room, MCR and Senior Common Room of life is still prevalent. If you declare an Act of Independence, is that not for every one? Often, it isn't: we can look back with hindsight and say this.

So how did this individual perform so well and yet so badly? Politically adept and yet morally corrupt? Well, in his day Jefferson was allowed to get away with what he did, in that slavery was seen as admissible, but with hindsight we can say that he was morally in the wrong. Even without hindsight, there were protestors and demonstrators back then. To complete the series of dates, regarding slavery, there was an International Slavery Convention passed in 1926 by the League of Nations, the forerunner of the United Nations, which provided for the complete abolition of all forms of slavery. This was reaffirmed by the United Nations in 1948 with the Universal

Declaration of Human Rights. Sadly, slavery has not disappeared, even now, in the twenty-first century.

Slavery has a long history. In fact, it goes back as far as historical records exist and there is no reason to believe it did not start long before that. In the ancient world all known civilizations used slaves, indeed they were essential for the economy. They were used in all areas of labour from the domestic scene to large-scale construction projects, such as the pyramids and palaces in Egypt. The ancient civilizations of South America also used slaves, but mainly in agriculture and warfare. It was quite normal for prisoners of war to be used as slaves, right up to the Greek and Roman empires. The Romans had a greater control over their slaves than most, having the power of life and death. In Rome it was not unusual for debtors to sell either themselves or their families into slavery to pay off their debts.

The first European nation to introduce slaves was Portugal. They had serious problems with a supply of agricultural workers, so the use of slaves is recorded as starting in 1444 and by 1460 between 700 and 800 slaves were being imported every year. These mainly came from Portuguese trading posts on the African coast. These were mainly Africans, caught by Africans, and then transported to the coast where they were sold. In Spanish South America the original slaves were the native peoples, but due to introduced disease and the nature of the labour it was perceived by the invading Spanish that the native South Americans were not as robust as the African slaves that the Spanish already had dealt with. This caused the Spanish authorities to start importing Africans as slaves, with a mighty death toll in transit between the two continents.

Generally speaking, slavery was commonplace until the

nineteenth century. It was during the eighteenth century that it was rumoured that Thomas Jefferson had a child with a slave by the name of Sally Hemings. It is this question that DNA can help to answer, although as we will see, when using DNA analysis retrospectively the results are never clear-cut. Sally Hemings was known to have had a white father and a half-white mother. This is significant in many respects. There is a word used for such a person, not often used now: a quadroon. It would seem to have originally been used in 1707 to describe a person who was regarded as a quarter black. This is extraordinary, because as a result of her parentage, both her peers and the grandson of Thomas Jefferson described Sally as nearly white. Whatever colour she may have been, she was thought of as very beautiful. Is it possible that she was a wholly owned servant, rather than a slave? Is there a difference? Whatever the truth of this, it is unlikely that we will ever know for certain.

Sally Hemings had been a very important and trusted member of the Jefferson household, working consistently for Martha Jefferson, wife of Thomas, until her death in 1782. Two years later, in 1784, Thomas had been appointed as minister to France to assist Benjamin Franklin and John Adams in negotiating trade agreements with various European states. This was a position that he held until 1789. It is during this period that his supposed tryst with Sally Hemings took place.

It was in 1787 that Mary Jefferson, daughter of Thomas, was sent to Paris to join her father. She was nine years old. This seems to have been motivated by the death of his youngest daughter, Lucy, of whooping cough. What was so extraordinary was the person who accompanied her as companion and chaperone: 14-year-old Sally Hemings. She was obviously a very independent adolescent, since she was thought capable of undertaking such a journey and looking after a nine year old whilst travelling 3000 miles across the Atlantic ocean on a sailing ship. This was against the specific instructions of Thomas, who wanted an older woman to accompany Mary. This was quite understandable, since sending a fourteen year old to supervise a nine year old on a transatlantic crossing *would* seem rash. This was 1787, and the voyage took the best part of six weeks. By the middle of the next century a change in the geometry of the prow of ships had created the clipper. One clipper, the *James Baines*, managed to cross from Boston to Liverpool in 12 days 6 hours. For Mary and Sally, however, the journey was much longer.

The point of disembarkation for the two girls was London. Here, John Adams and his wife, Abigail, met them. The captain of the ship suggested that Sally Hemings went back to Virginia, but since she was now in Europe it was decided that she should stay with Mary. Thomas Jefferson did not come to personally escort the two girls to France from London but sent an assistant, who took the pair to Paris and the official residence. It was here that Sally's brother was being trained as a chef. He had come across with Thomas Jefferson, and like any slave of the day he had arrived with his master with no ability to speak French upon his arrival. By the time that Sally arrived he had been in the country for the best part of three years and he had no doubt learnt a considerable amount of the language. He had, in fact, hired a French tutor for himself.

The service staff at the residence was essentially complete by the time that Sally and Mary arrived at the official residence. Mary's sister, Martha, was already there and so in no time at all Mary and Martha were both sent to an exclusive Catholic boarding school. It is of interest to note that Sally Hemings stayed with Jefferson. She could have quite easily left because although she possibly thought there was nowhere for her to go, as far as the French government was concerned she was free, and not a slave while on French soil. Admittedly the level of comfort must also have constituted quite a bond to Jefferson. The idea of keeping up appearances was around then just as it is now, especially in a fashionable city like Paris, and even more so for a senior diplomat like Jefferson. To this end Thomas Jefferson spent considerable amounts on his household. For a senior figure representing a foreign country it was seen as important that not only the head of the household should be well turned out, but the rest of the household should be as well. Badly dressed servants would reflect badly on both Jefferson and the USA.

It was in 1789 that Jefferson, his two daughters, Mary and Martha, as well as James Hemings and Sally Hemings, set off across the Atlantic to return to Virginia. Contemporary narratives suggest that Sally gave birth not long after returning from France, which implies conception having taken place sometime before leaving Paris. The child who was supposedly born was Thomas Hemings, or as he is sometimes known, Thomas Woodson. The story, however, does not stop there.

Sally Hemings went on to have a further six children, William (born 1798), Harriet (born 1801), James Madison (born 1805), Eston (born 1808) and two other children, both of whom died in infancy. Thomas Jefferson supposedly fathered all of these offspring. Although not conclusive in any way, it is of interest to note that William Hemings ran away from the home of Thomas Jefferson, Monticello, in 1822 when he was 24 years old and went to Washington where he lived as a white man. Similarly

Harriet Hemings left Monticello in 1822, aged 21, with the knowledge and help of Thomas Jefferson and went to Philadelphia where she was accepted as white and married into a white family. Indeed, it is said that by 1873 all of the children of Sally Hemings had more or less joined the white community. We know that Sally was fair skinned, but it is reasonable to assume that her children must have been of even paler hue for two of them to pass as white. Rumours and quiet whispers started at this point regarding the relationship between Jefferson and Sally Hemings. At the time it would probably have been scandalous had it been publicly acknowledged, but privately such liaisons were not uncommon.

According to Madison Hemings it was Sally's grandfather who gave the family the name Hemings. He was a captain of an English trading vessel that sailed between England and Williamsburg, Virginia. Further, Madison claimed that his great-grandmother was the property of John Wayles, who would not sell the Hemings child to him but kept and raised her himself, later to become his concubine. All of the children of this liaison carried the name Hemings. After the death of John Wayles the Hemings family passed to Martha Jefferson. It is interesting to note that Madison also went on to say that by the time Thomas Jefferson left France Sally had gained a considerable grasp of the French language and was extremely reluctant to return to America. In France she was free but if she went back to America she would once again be a slave. To persuade her to return Jefferson made pledges of extreme privileges and also promised that her children would be freed at the age of 21. At the same time Sally stayed with Jefferson until his death in 1826.

It might seem odd that Sally Hemings was not formally freed from slavery in the will of Thomas Jefferson. It was a far more informal event than this; she just went to live with her sons, Madison and Eston. It was probably thought that at this stage, with the man dead, there would be no point in stirring up rumour, so Sally, already middle-aged herself, simply left Monticello.

An explanation for the Sally and Thomas liaison is that Jefferson was a big figure in politics; a powerful man (and power itself) can be very attractive. The rumours that Thomas Jefferson had Sally as both slave and lover came into the public domain when Jefferson started his first term as President. This first term was a controversial result because the two Republican candidates, Jefferson and Aaron Burr, received equal votes so it was up to the House of Representatives to decide which one was to be President and which one Vice-President. Jefferson won the day and took office on 18 March 1801, although the election had taken place the previous year. In 1804 he was re-elected for a second term. During his first term a journalist, James Callender, published a series of attacks on the President that stated that he was the father of the children of Sally Hemings. It later transpired that James Madison Hemings told a newspaper in 1873 that his mother had confided in him that his father was indeed Thomas Jefferson. There might be all manner of personal and political reasons for these suggestions and accusations. In the early nineteenth century, these things were seen as scandalous. What it does indicate is that at this stage it is all just a matter of opinion. Basically there is little by way of evidence either way, just garbled reports and circumstantial evidence from published reports of the time. What is required is something that could push the story in one direction or the other.

This brings us to the point where DNA analysis may be able to clear the air. The reasoning is that if living relatives could be found that could be reliably documented as being related to Sally Hemings, but through a paternal line (her sons, their sons and their sons down to the living male descendants), it may be possible to compare the male Y chromosome with known living descendants of the Jefferson line. The Jefferson lineage goes back two generations further than the Hemings, the great grandfather of Jefferson having had two male children, one of which was the father of Thomas and one of which was Field Jefferson, from whom the living Jefferson DNA would come. This does rather assume a level of fidelity over the generations, but as we shall see it seems to have all worked out.

It is not just Thomas Jefferson that is thought might have been the father of Sally Hemings' children. It has also been suggested that one, or both, of two brothers might have been responsible. These were Peter and Samuel Carr, nephews of Thomas Jefferson. They were related maternally to Thomas Jefferson, rather than paternally. This is an important point because it means that they would have inherited the male sex-determining chromosome, the Y chromosome, from their father and it would be expected to be subtly different to the Y chromosome of Thomas Jefferson.

In this investigation it is the Y chromosome which is of major importance because although over time it will pick up mutations of various sorts, it is passed lineally down the male line, so all male descendants of an individual will have essentially the same Y chromosome. The Y chromosome is so called not, as is commonly thought, because of its shape, but for a very quirky historical reason. When the X chromosome was first seen in 1856, by Edmund Wilson, it was seemed mysterious and enigmatic, so what better than to call it as if unknown – X. Following on from this when the Y chromosome was recognized it seemed logical to give it the next letter of the alphabet. The Y chromosome differs from the X chromosome in many very significant ways. The X chromosome is large and carries very many genes that are of particular importance, in that it is not possible to survive without an X chromosome. Many of the genes can cause severe problems if they are damaged, such as Duchenne muscular dystrophy. In contrast to this the Y chromosome is one of the very smallest of the human chromosomes and is virtually inert. It contains only a single functional gene called *Sry*, which starts the cascade of genetic activity resulting in a male rather than a female. There are other non-functioning genes while the rest of the DNA of Y chromosomes is non-coding, although it is probably important in cell control.

Among the non-coding regions of the Y chromosome there are measurable areas that do not vary very much, if at all, from generation to generation. So if we go back far enough we can find a common ancestor of two different family lines: one that went on to give President Jefferson and then, possibly, Eston Hemings and the other line which gave Field Jefferson and his descendants. This can be seen in the abbreviated family tree (Figure 7).

These two lines diverged after President Jefferson's grandfather, also Thomas Jefferson. Consequently, after several generations these two lines now have living descendants. If they all share a common ancestor then they should all share the same Y chromosome.

What was found was that there was a small mutation in one of the Y chromosomes tested, but some changes would be expected in such a large number of individuals through so many generations. All the other Y chromosomes, however, matched. Now, like all DNA analyses, if results

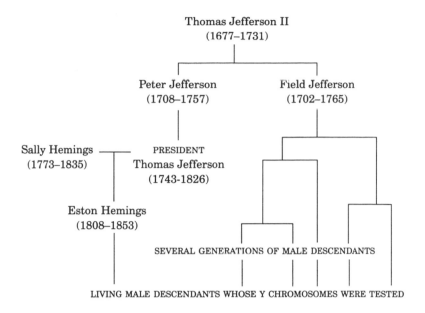

Fig. 7 The Jefferson family tree

The potential relationship between Thomas Jefferson, third President of the USA, and Sally Hemings. By looking at the Y chromosome it can be seen that living descendants of Eston Hemings should have the same Y chromosome structure as descendants of Field Jefferson if they are related

do not match then we can be sure that there is no relationship. On the other hand, as in this case, if there is a match, we cannot say with absolute certainty that it proves a relationship. What we have to do is weigh in the balance the prior odds. Eston was Sally Hemings' son, Sally was a member of the Jefferson household for most of her life and there was never any suggestion by her that it was anyone other than Thomas Jefferson that was the father of her children. Besides this, even though Sally Hemings was of very pale skin, some of her children moved, with no apparent problems, into white society, which by implication would indicate a white father.

There was a suggestion, primarily made by one of Thomas Jefferson's grandsons, Colonel T. J. Randolph, that Sally Hemings had been the mistress of Peter Carr, whom Jefferson had brought up at his house, Monticello. Peter Carr was a nephew of Thomas Jefferson, being the son of his sister and her husband Dabney Carr. Now, because the chromosome which was looked at was the male, Y, chromosome, it would be expected that if Peter Carr was the father of Eston Hemings and his siblings, the descendants of Hemings and Carr would have the same Y chromosome polymorphisms. This is because although the mother of Peter Carr was the sister of Thomas Jefferson, he would have inherited his Y chromosome from his father. The Y chromosome found in the descendants of Eston Hemings was not compatible with having come from the Carr line of descent.

So we do not know for sure whether Thomas Jefferson was the father of the children of Sally Hemings: it will probably never be known for certain, but DNA has given us a bit of a pointer to back up the other documentation. DNA can only give an absolute answer when it gives exclusion, and that is not what we have here. What we do have is a piece of a jigsaw which does fit quite nicely with the story, but we can never be certain and we will never see this particular jigsaw completed.

Using Y chromosomes can help in all manner of ways when it comes to genealogical searches by individuals from less famous families than Thomas Jefferson. In January 2002 the Public Records Office in the UK made available online the 1901 census of Great Britain. What they had predicted was a hit rate of about 1.2 million visitors a day and this is what they had planned for. What they actually got was 7 million hits a day. The system was so overloaded by this that it had to be temporarily closed down. This reflects the passion people have for their past. People want to know where they come from and who their ancestors were. My family, for example, originated in and around Much Wenlock, a small village in Shropshire, England. It is undeniable that ancestry holds a fascination for most people. Tracing relatives, as we have seen with the Hemings and Jeffersons, does rather depend upon a presumption that there has never been any infidelity in the family. In England the earliest National Register of births, marriages and deaths started in 1837, with other records tending to finish in the middle of the seventeenth century because of the English Civil War that started in 1642. On a real time scale this is not such a long way back. By bringing DNA analysis of the Y chromosome into the picture it may sometimes help take the story much further back into the past.

There are broadly two ways in which Y chromosomes can be used. The first we have already seen, depending on the way that direct descent through a male line results in inheritance of identical Y chromosomes. The second way it can be used is by knowing the average mutation rate of the tested Y chromosome sequences. By knowing this it becomes possible to tell how long ago in the past two individuals with different Y chromosomes shared a common ancestor.

One rather interesting case of using Y chromosome DNA also involves tracking surnames, which, like Y chromosomes, travel through the male line. Brian Sykes of the University of Oxford carried out this investigation, which involved the family name of Dyson. This particular name was always assumed to refer to the work that the original Dyson carried out – something to do with dyeing. It was common practice when family names were first taken in the Middle Ages for the name to reflect the type of work, such as Smith or Mason, or local features, such as Lake or Hill. All manner of records were looked at to try to determine when the name originated in Yorkshire. These included such unusual sources as returns from the Hearth Tax. The Hearth Tax was introduced into England in 1662 and repealed in 1689, having raised £170,000 every year. It was levied on all houses unless the occupier was exempt from church or poor rates, or was living in a tenement under the value of £1 per annum, and not having land or goods of greater than £10 in value. Luckily, it was possible to go back even further; in fact it was possible to find a record of a manorial court case from 1316. The manorial court was an interesting institution, with officials of the manor holding office and deciding any necessary penalties or sentence, yet holding no sway over the law itself. It was the tenants of the manor who were expected to attend the court and give judgement on the case. That the records from 1316 still exist is a testament to the constitutional inability of civil servants to dispose of any records!

This particular manorial court was looking into a charge of stealing cattle from a neighbour's land. Those who were being prosecuted were John Dyson and his mother Dionisia, normally referred to as Dye, implying that John Dyson was not a dyer, but Dye's son. Samples from as many Dysons as could be found: were tested, by analysing the DNA of the Y chromosome. Perhaps surprisingly, a very positive result was found, 85% of the male relatives shared the same Y chromosome. By looking at the mutation rate and calculating backwards, Brian Sykes showed that the ancestral chromosome had indeed originated around the beginning of the fourteenth century.

Great Britain has a very mixed population, with Angles, Saxons, Romans, Normans and Vikings in the gene pool. This is to name only a few of the large and small groups that have come to Great Britain over the last few thousand years and stayed to add to our diverse gene pool. There has been mass settlement, mass invasion and large-scale immigration. These groups left a legacy of names and genes about which there is an intense curiosity reflected in the modern obsession with genealogy. So far it is the Viking connection that has been mapped out most accurately. By comparing current Y chromosomes with a panel of Y chromosomes from supposed Viking descendants in Scandinavia, it is possible to determine whether or not your ancestors were Vikings or not.

All of these processes of looking into the far distance past using DNA are fraught with potential problems, perhaps the greatest of which being the assumption that results of DNA analysis are infallible. In any analysis there will be room for residual doubt, but when it comes to these attempts to look back into the very distant past great care needs to be exercised in interpreting the results as absolute truth – we can never know for sure what the truth is.

6 ANCIENT DNA

There is a sad trend in human behaviour where families behave in a manner that is not always entirely pleasant. People often behave unpleasantly. They lie, cheat, steal and play foul in love and war. However, this instance refers to the not infrequent case in which an individual, in reality possibly long neglected, but sometimes not, dies and long-lost relatives appear apparently out of nowhere to claim an inheritance. In fiction it is usually a vast estate complete with baronial hall where an individual turns up who claims to be a long-lost son. He then assumes the mantle of this elderly father, restores law and order and upholds the noble family tradition, banishing the pretenders to the proverbial throne. That is the romantic fictional situation; reality is rather different.

In reality often what happens follows a very shabby path. It runs along the lines of an old man dies, perhaps alone, perhaps with a companion or wife. Most importantly, no will is left so the estate has to go to probate – that is official verification of a will, or more likely if there is no will to try and decide what the estate is worth. With no will the next of kin effectively gets the lot, after probate tax, or death duties; if no relatives can be located the estate goes to the state. So we have a potential situation where a neglected individual dies with no will and no known descendants. What would normally happen is the exchequer would benefit, but there is a slight catch. If a will after probate is greater than £5000, the will becomes part of the public domain, although it does not officially have to be published. If it is far greater than this amount and there are no known relatives it becomes ever more likely that disputes will arise. There is money to be made, potentially large amounts of money from an intestate death; all you have to do is prove a relationship to the deceased. But without some good documentary proof this is difficult – unless you can get a DNA sample. As we have seen with the Romanovs, this is not entirely impossible, but it is difficult.

If we are to find out whether an individual that has come forward after a will has been made public is really related to the deceased, it is important to find out one thing first – was the individual buried or cremated? This is almost bound to have already taken place by the time all the details of probate have been sorted out. It may sound like a silly question, but the fact remains that unless it is asked it is possible to be quite a long way down the road towards organizing DNA analysis before you find out that the deceased was cremated and it is realized that an urn full of ashes contains no DNA. In the case of a buried body the amount of residual DNA that is available depends on many different factors, such as the water content of the ground, the acidity of the ground, the temperature and the depth of burial. As we shall see in the next chapter, the state of decay of a body can tell us a lot about what has happened to a body since death. A good indication of how well preserved a buried body can survive can be seen in bodies that have been discovered in acid bogs both in the UK and Europe where after a thousand years there was still skin present on them, effectively preserved by a natural process of tanning in the brown acid waters of the bog.

In the course of a normal burial process decomposition will eventually take place completely. In fact under certain circumstances, even the bones will disintegrate, as the calcium is leached from them, leaving only biologically decomposable cartilage. There is an urban myth about buried bodies, which apparently started in the nineteenth century: that hair and nails continue to grow after death. To put this to bed and lay your mind at ease, this is simply not true.

One such case where it was necessary to investigate a buried body was associated with an individual who had died in his 60s, with no known relatives and apparently of ordinary means. He left no will; he was a bachelor with no brothers or sisters and both his parents were dead. When the size of the estate was discovered, which was based upon property, a girlfriend from the past of the dead man came forward insisting that her son had been fathered by the deceased, making him, if it was true, the sole inheritor of the estate. It seemed the only way to clear up the situation was to compare DNA from the child with the mother and the deceased, to see if it was at least possible that he was the father. A court order was obtained to exhume the body. This is not so easy as one may think, rules are very strict – as they should be – after all you would not like to think that you or your relatives could be dug up on a whim. A few notes on the process of exhumation will show this.

If a body is buried in consecrated ground it is protected by the ecclesiastical court; if it is buried in ground which is not consecrated it is the responsibility of the civil courts. If a body is buried in consecrated ground, then removing or disturbing the body is both an ecclesiastical crime and criminal offence, which would result in an arrest and criminal record. In the USA this would be regarded as a misdemeanour, which is a lesser offence than a felony, such as murder. Even when not on consecrated ground disturbing an interred body is still an offence and it does not matter what the motivation is: without official sanction this is illegal. Very persuasive arguments have to be put forward to the court if you are brought up on the charge of illegally exhuming a body.

There are basically three reasons for exhumation. The first is when, for example, a body has to be moved for sanitary reasons or perhaps when the relatives request a move from one consecrated site to another. It also includes the case where a road needs to be widened and cuts into a graveyard. The second case for exhumation requires a licence from the Home Secretary, if it is either desired or necessary to remove a body from one unconsecrated site of burial to another site, either consecrated or unconsecrated. The third broad reason for exhumation is by the order of the coroner when there is a sufficient suspicion of a crime to justify such a radical activity, or when the outcome of a lawsuit regarding an intestate individual cannot be solved in any other way. In the USA the situation is slightly different. If removal of a body from one place to another is wanted by relatives, they apply to the local board of health, but if there is a dispute among relatives over this it is for a court to decide whether it should be carried out, or not.

In the particular case of our intestate gentleman an exhumation order was made because there seemed no other way of finding out if the child was related to the deceased. It would not have been necessary to disturb the dead had there been living siblings, or living parents of the deceased, because it is possible to re-create a genetic family tree without all the intervening generations. It would have been possible to surmise what the DNA profile of the deceased was from parents and siblings, and therefore whether the child could have been fathered by the dead man, but no such thing was possible in this case. Even if it had been possible that way to show that there was a potential relationship, it would still have been a matter of calculating a probability. The only way of being sure whether the deceased was the father of the child was to exhume the body and hope that sufficient DNA was available, such that an STR profile could be produced. One thing was certain, it would have to be an STR profile, because the body had been buried for two years before exhumation, and although in a coffin, decomposition starts quite quickly from within due to the presence of bacteria and fungi already on and in the body.

When dug up the body was unrecognizable. It was not even possible to determine the sex of the corpse; after two years, decomposition was well advanced. The exhumation was carried out in a sensitive way, but there was a problem. The original arrangement was to take the body to a mortuary, but permission for this was withdrawn at the last moment. As a consequence it was necessary to take the body to a local undertaker where samples were eventually taken. A pathologist using a saw took these samples from one of the thigh bones. There was always a risk that there would be contamination of the samples from material on the saw originating from previous use. Although this is potentially a very real problem, in this case it did not seem to have been a problem with this investigation because of the standards of cleanliness that the pathologist insisted was applied to the instruments. The reason that contamination can be such a problem is due to the extraordinary sensitivity of the PCR reaction. Even so it was originally thought unlikely that any useable DNA would be extracted which would be suitable for amplification using PCR. It was a thigh bone that was chosen because these long bones, as they are called, are the site in which blood cells are made, the bone marrow. Being in the centre of the bone the cells, and therefore the DNA, are relatively well protected, certainly far better than the DNA found in soft tissue, which is the first material to decay completely.

The bone samples were put into sealed bags and taken to a Home Office Forensic Science Service laboratory where the analysis was to be carried out. The first laboratory failed to produce any sort of a result, so the remaining samples were forwarded to another Home Office laboratory, where technicians more experienced in these difficult DNA extractions and analyses would try again to produce a result. This was more successful: a result of sorts was produced. At this time it was still early days for STR profiles, so whereas now ten different STRs are used to produce a profile, at that time only four were used. This reduces the discriminatory power of the profile quite considerably, but in this case that was not going to be relevant.

As with all DNA profiling, if two profiles match, or are related, a probability has to be generated as to whether the profile match by chance or not. If there is exclusion, that is, the profiles do not match or appear to rule out a blood relationship, then no amount of arguing can change it. In paternity disputes such as this, the situation, as we shall see, is slightly different. In a case of this sort, by having a blood sample from the mother and child, it can be worked out what the profile of the father should be. Taking the relevant part of the mother's profile away from the child's does this. What is left must have originated from the father. By repeating the experiment several times a consistent result was found which gave a match on three of the STRs and a mismatch on the fourth. Normally it would be automatically assumed that this ruled out the dead man as the father of the child and indeed that is what the High Court decided. But in these sorts of cases it is not so straightforward.

When a sample from a scene of a crime is compared with a sample from a suspect, exclusion is exclusion. Paternity is slightly different because it is during the production of sperm and eggs that mutations take place, so a child may not necessarily have the expected profile from the father, a mutation having taken place which altered the number of repeats in an STR. It has to be said that this is not a common finding but single STR exclusions, such as in this case, would now warrant further testing to clarify the situation. Using a larger panel of STRs would do this, and at least two would have to mismatch before there would be any certainty that it was a genuine exclusion. In these situations, the more STRs that are used the better. This is because if there is no relationship, the more that this result would be seen as exclusion.

Another exhumation case was one of the very first to have been conducted using PCR in the USA. It was in 1986 that the case of Pennsylvania vs. Pestinikas took place. Mr and Mrs Pestinikas were the operators of a rest home in Pennsylvania where a very old man had apparently died of starvation. As a consequence Mr and Mrs Pestinikas were charged with negligent homicide. During the trial the defence lawyers questioned the credentials of the physician who had carried out the autopsy on the man. It was this physician that had given the cause of death as starvation. Following this the judge ordered that the body be exhumed for another autopsy. Since he had not been buried long, there was adequate soft tissue for extraction and analysis of DNA. The story becomes a little more convoluted here because after the first autopsy, but before burial, the body had been in the custody of the defendants, which made the physician carrying out the second autopsy suspicious that the internal organs had been switched with another body to hide the true cause of death. So at this point it became more than just an autopsy: it was necessary to find out if the organs in the body really belonged to it.

To test this hypothesis that organs had been switched, the Cetus Corporation became involved as DNA profilers. They managed to extract sufficient DNA from the internal organs and other tissues, such as muscle, which obviously belonged to the body, so that they could be compared. In this case all the profiles matched, so although they, the defendants, were found guilty of negligent homicide, they were acquitted of the charge of tampering with the body.

So, outside the arena of retrospective criminal and paternity cases, what is it that we can find from DNA that might tell us something about our past as the species *Homo sapiens*? Well, the answer is perhaps a little surprising, because it is 'quite a lot'. For one thing, properly mummified remains retain quite a lot of DNA. It follows a simple set of rules: if the tissues have not decayed there is probably DNA available for analysis. This reflects the intrinsic robustness of DNA, which stands to reason since not only is DNA in the heart of the cell, but some of these cells are in the heart of very solid structures, like bones. After all, you would not want your genetic material mutating every time there was some sort of environmental assault on it. If that happened it would spell disaster for the organism. What you want is a robust

molecule taking your genetic information across generations as unaltered as possible. Mutations do occur, of
course, but the basic molecule is pretty solid. Such is the
power of this resistance to decay in dried and cured or
mummified tissue that such animals as the woolly
mammoth or the quagga have rendered material suitable
for analysis. In the case of the mammoth it was from a
corpse found in the permafrost of Siberia, deep-frozen for
thousands of years. In the case of the quagga it was
slightly different.

There are a lot of skins of quaggas in the natural
history museums of the world. Quaggas were a distinct
species of zebra, basically a brown animal, with a white
tail and legs, but with the traditional black and white
stripes of the zebra on its neck and head. It is now extinct,
largely, if not totally, due to hunting by humans. The
Boers hunted this once common animal for its skin, which
was sturdy and strong. So, long after the once large herds
had been decimated, the skins were still in routine use,
even after the last quagga was seen in 1883 and finally
declared extinct. Even though they were hunted to extinc-
tion, the DNA of the quagga remained in the cured skins.
It has been suggested that it may be possible to extract
sufficient DNA from a preserved skin to clone the quagga
and resurrect this extinct species. This is most unlikely,
for even if all the DNA is present, it will most certainly be
broken and damaged. With STR analysis only very short
pieces of DNA are necessary for the construction of a
DNA profile, but to clone an individual organism you not
only need all the DNA, but it has to be in the right order
and wrapped up with the right proteins, so that they can
actually function. It is not enough to have a DNA
sequence: control of gene expression and how it is finally
expressed depends upon so much more than just a

sequence of bases. As an analogy, I could give you a dictionary and say 'all the words of Shakespeare are here – you just have to get them in the right order and with the right punctuation'. Just knowing the alphabet is not enough.

In forensic applications of DNA profiling using partially decayed material, we do not need a complete genome, that is, all the DNA in the right order. What we need are DNA fragments, large enough and from enough cells, so that we can produce a reliable DNA profile. This extends to mummified remains. Perhaps we can determine the relationship between populations and possibly even between races, using these DNA sequences. This is the history within our genes. It is staggering just how much information about our individual relationships can be gained from our DNA.

Before proceeding further it is worth considering some of the least likely scenarios regarding ancient DNA that have gained common acceptance. Perhaps the commonest of these is the ideas encapsulated in *Jurassic Park,* filmed in 1993. Even if it were possible to extract intact DNA from an insect that had bitten a dinosaur, another problem would arise as to embryology. As we have seen, it is axiomatic that just knowing a DNA sequence does not give an insight into the growth and development of an embryo. There are fundamental problems with the idea that extracted DNA can be used for cloning. First, of course, is the difficulty of finding complete DNA. Unless the DNA is complete and in the right order it will not work correctly. This is no simple point; DNA is sorted into chromosomes and when the DNA is rearranged all manner of disasters lurk. This is the commonest cause of cancer, the most well-known example being in human leukaemia. Attempts were made using PCR to detect insect DNA from insects which were trapped in African amber, just as in *Jurassic Park*. Some results were produced, but on detailed analysis of the DNA products they turned out to be broadly from three origins, mammalian DNA, fungal DNA and self-replicating PCR products. This in itself was a disappointment, but when closer investigations were made of the proteins of the insect, another observation emerged. Even though the insect looked intact, with lots of structural details on the outside, chemically all of the proteins that were extracted and tested were found to be degraded. No chance of dinosaur DNA here then. It also shows that even with immense care, the use of PCR can generate artefacts from the slightest contamination.

On the other hand, to produce a DNA profile based on STR analysis, DNA can be broken and damaged, as long as there is an intact section of material that contains the STR to be measured. Since it is generally only a sequence of approximately 100 base pairs in length that is required, the probability of a break occurring in that particular sequence, out of the approximately 3000 million bases present in the human genome, is very small. It is well understood, now, that even under the most favourable circumstances it is unlikely that DNA could survive for more than 50,000 or 100,000 years. Remember that even an extreme old age is only 100 years and that the history of the Christian church is only 2000 years; this pales into insignificance when compared with 50,000 years, or even more with twice that age.

There *is* a possibility that ancient DNA, at least ancient in human terms, can be extracted and analysed. Not with a view to cloning a Neanderthal, say, but producing a DNA profile which allows us to put an individual into an historic context. This is the search for ancient human

DNA. This is the search for material that, quite simply, makes genetics the most exciting subject in the world. It can reach backwards, it can reach forwards through generations, and it truly is the most amazing subject.

But how far can we go, either forwards or backwards? The answer is not simple and not complete. It is not simple because we cannot say that this is all we will ever know about life and the preservation of DNA. At one point, in 1992, it was suggested that a PCR result had been produced from a bone of *Tyrannosaurus rex*, which was 65 million years old. Of course, this means that the bones were not Jurassic, but then neither was *T. Rex* – strictly speaking this dinosaur was Cretaceous, which came after the Jurassic period. Unfortunately the DNA results were unusable: basically they did not work. One of the most recent animal remains that have been used for DNA analysis is that of frozen woolly mammoths from Siberia. Having been in permafrost for more that 50,000 years (bearing in mind that is twenty-five times longer than our current Christian calendar is dated), some sort of result was produced. There was a more important result from this than just that, however. It has become apparent since then that it should be possible to hunt DNA backwards from bone which has been either dried or frozen for anything up to 100,000 years.

Moving on, then, just as you might want to know where your parents came from, so we all have a desire to know where our families arose. This might be possible in the short term, but when we go further back, relationships between families become progressively more difficult to identify. This becomes especially more so as the distance of time muddies the water. This works as a neat analogy for the way that DNA evidence is muddied. We may want to know where *we* as a race originated. With family locations and genealogies, eventually we get to a state where there are no reliable documents, even no documents at all, regarding a specific family.

The origin of man, too, is shrouded in almost undocumented mist. We are unable to look back in time, except through our molecular past and our bones and fossils. It is thought that humans started as prosimians, relatively unsophisticated generalists that occupied a wide range of different environments. Although they did almost certainly had five fingers they certainly did not have opposable thumbs, that is, a thumb tip which can be touched by every other fingertip on the same hand. As the climate changed, so did the vegetation, and so did the type of species that flourished.

As time moved on we come to the Australopithecines, that lived about 5 million years ago. This is still too far in the past for us to use DNA to investigate, but given the archaeological data, anthropologists seemed to be on the right track. As time comes closer to that standard of less than 100,000 years, so the possibility of DNA as an archaeological tool becomes ever more possible. It seems that we are a recognizable sub-species of *Homo sapiens*. In fact it is probably the case that Neanderthal man could be classified as *Homo sapiens neanderthalensis*. About 30,000 years ago, a sub-species appeared, *Homo sapiens sapiens*. This is me and you, so far. As with all of these things, it must be remembered that there was not one species A which suddenly turned into species B. It was a gradual stage without any marked delineations. Just as a son or daughter isn't hugely different from their parents these days, so it was the same story all those years ago. It just has to be borne in mind the huge time scales involved here.

It would seem that modern man appeared about 30,000 years ago, which puts us in the time frame of extracting

DNA from old bones. What could stop us? Well there is always the problem of contamination. Sample contamination is a real and very important problem. A simple and easily given demonstration of contamination is shown by the pores in bones that will eventually allow the entrance of foreign material. It is just the same as when a bone is cut: foreign material can be introduced in all sorts of different forms, such as skin flakes, like dandruff, from the scientist doing the analysis, or from the tools used. Skin flakes form the major part of household dust and laboratories that introduce PCR into traditional laboratories often find that benches and other static equipment are contaminated with sufficient DNA to give false results in PCR reactions. This reflects the staggering sensitivity of PCR analysis.

There are some truly amazing demonstrations of the sensitivity of DNA, some from forensic applications and some from medical research. It is possible to use a technique called cell sorting to separate individual cells from each other. This utilizes the charge that can be induced on the surface of cells in an electrical field. As the cells pass down a nozzle a single cell is encapsulated in a single droplet and can be diverted into a container by putting an opposite charge across the gap through which the droplet passes. In this way it is even possible to separate single sperm from each other, which can then have its DNA analysed. This is in many ways the ultimate sensitivity, because not only is a sperm a single cell, it is a single cell with only half the DNA content of a normal cell. In forensic applications the level of sensitivity does not involve single cells separated in a laboratory, rather, it involves finding cells in very difficult circumstances. Some of the most spectacular successes in forensic science involve very delicate and intricate processes to recover cells from some most unexpected sources.

Probably the most obvious subjects for analysis are cigarette ends. This was demonstrated as possible very early in the history of STR analysis. The reason for this is that it is quite commonplace for cells from lips to come off and stick to the cigarette end, whether it is filtered or not. The problems with taking a DNA profile from a cigarette end and relating it to an individual are twofold. The first potential problem is a shared cigarette, which would result in a mixed profile that could be interpreted in any number of ways. As a result of this the analysis may be invalid. The second problem with mobile evidence, such as a cigarette end, is that it may be associated with an individual, who may well have smoked it, but it is literally impossible to tell from a DNA profile where the cigarette was smoked. It is always possible that a malicious individual could leave someone else's mobile evidence at the scene of a burglary.

A rather more unlikely source of DNA for analysis is taken from the back of stamps and envelope flaps. We might immediately sit up and think that envelope flaps, just like cigarette butts, are mobile evidence, but the difference is based on the reason why you would want to recover DNA from an envelope or the back of a stamp. If someone receives a poison pen letter, a threatening letter, or a blackmail demand, the DNA from a licked envelope flap or stamp could link the letter to a suspected individual. This is extremely difficult, working at the very limits of the technology, but if saliva has been used to stick the envelope down, there will be cells there and the cells will contain DNA. This may become progressively less important with the increased use of self-seal envelopes and self-sticking stamps.

Once the techniques became reliable and repeatable, DNA was gradually extracted from a variety of material, some of which was really quite unusual. There were 300-year-old bones from an English Civil War cemetery. Another set of bones from a Judean cave were dated at about 7,500 years old. Both of these sets of bones, the Civil War and Judean cave bones, rendered enough DNA to produce a PCR result. These were not analysed using STRs, but by looking at specific genes, with the aim of determining the range of variation over time and among different populations. The reason for this approach is that it lends itself to analysis of other material than just human. A good example of this was when a mammoth tooth was obtained for analysis and sufficient DNA was extracted for it to be analysed. Before an STR profile can be constructed we would need to know the sequence either side of the target sequence.

The common origin of life can be inferred from the incredibly conservative way in which some of our most fundamental gene products – proteins – are carried through evolutionary time so unchanged that we know that the gene from, say, a plant is identical to that self-same gene found in humans. It is reasonable to assume that most, if not all, animals will have the same gene sequence. It does not have to be exactly the same – there is, after all, some flexibility in the DNA code, but broadly it will be the same. So if we look at the same gene product, we can be fairly sure that the gene sequence must be close between the compared species. This can be looked at in terms of thinking that a woolly mammoth is a big hairy elephant. As a consequence they share a wide range of enzymes and protein products. Also, they consequently have to share the same genes; there are only so many genetic combinations that will code for the same protein sequence.

It therefore becomes possible to look at an elephant (in

fact any mammal) and find a surprising consistency of protein structure and function. So we move from a simple genetic profile to something far more profound. It all of a sudden becomes possible to look backwards far deeper than any one had ever thought possible before. To make such investigations a careful choice regarding which gene to look for is of paramount importance, when looking at non-human material. This can then give a lot of information about the relationship between changes in specific genes and if enough different species are looked at it becomes possible to produce what is effectively a family tree of a range of closely related species.

With humans we already know the sequences that can be used to produce an STR profile. We also know enough about mitochondrial DNA to be able to use this to investigate human relationships, relationships between racial groups and, for example, where specific groups originated.

A simple question may be asked, which is 'where do we come from?' Of course an alternative question would be 'why should we want to know about the ancient history of our species?' The answer to this second question is simple: human curiosity. This is the natural curiosity about who we are and where we came from.

In the mid-nineteenth century, an event took place in a quarry above the Neander valley near Düsseldorf. Apparently even the quarry men digging there realized that although human-like, the skull fragments they dug up were not the same as their own skulls would be. This was 1856 and any small town, away from a large metropolis, normally had only two people to address intellectual questions to. The first might be the doctor and the second might be the teacher. In this particular case the quarry workers alerted the local schoolteacher. The schoolteacher recognized certain differences between modern humans and the skeleton that had been unearthed. Without the knowledge of evolution, which only appeared several years later, there was little chance of understanding what the true significance of these bones really was.

These bones were called Neanderthal, a term which is now eponymous with a subhuman form that predates our present human form. But do not imagine that Neanderthal man was in some way a direct antecedent of modern man. Again it comes down to a 'family tree' and here that phrase is in quotation marks for a very good reason.

Our common perception of a family tree of life is really the presumptive idea of Ernst Heinrich Philipp August Haeckel. Born in 1834 he was a great zoologist, but he has added two astonishingly incorrect notions to the common heritage of biology. The first is perhaps is not so well known. It is that ontogeny recapitulates phylogeny. This is a notion that as an embryo grows it goes through all the evolutionary stages of the taxonomic group. An example of this would be the development of a human foetus exhibiting what are apparently amphibian characteristics as it develops. It may look like it, but it would never 'make it' as an amphibian; evolution and development is far more complicated than that. Although this is appealing as an idea, we must not be fooled by simple explanations that skim the surface.

The second major problem that Haeckel has given to us is far more difficult to gainsay. This is the concept of the 'tree of life'. This, frankly bizarre, idea was that in some way we, humans, were at the top of some mythical tree of development, with bacteria and amoeba at the bottom. From the bottom of the tree it would then move up through various stages, such as fish, amphibians and retiles, culminating in apes with us, humans, at the top. This is not how evolution works. Evolution is neither linear nor predictable and we did not evolve from chimpanzees. All of the great apes, including chimpanzees, gorillas, urangutans and humans, had a common ancestor, but we are not related in a linear way. We, and those different species, did not evolve from each other. As a result, it would be of interest to find out exactly how closely we *are* related to our great ape relatives. For that to be answered we need to go back to the point at which all of the great apes became separate. There is a common ancestor, not a missing link. It still all presages the big question: where did humans come from?

The bones from Neanderthal man found in Germany provided ample resources for the production of a linear idea of humans evolving from a sort of half animal and half human. We now realise that this is simply not true. Somewhere there is a common ancestor of both species, but they did not arise one from the other; they were always independent and separate species.

At the time speculation about Neanderthal man was inevitably related to anatomical detail only. Anatomical detail at this time and for a long time after was based on opinion. This process is sometimes called Cuvier's bones. This is after the work of Georges Cuvier (1769-1832), a brilliant anatomist and the developer of the science of comparative anatomy. His knowledge was such that he could deduce the form and function of a complete animal from a few fossilized bones. Unfortunately this did not always work, at least by modern standards of anatomy, with our far more detailed knowledge of the animals that he was working with. However, there would be considerable *hubris* in suggesting we have got it entirely right. After all, how many of us have seen reproductions in colour of what a dinosaur looked like? In actual fact, we have absolutely no idea what sort of skin colour any dinosaur would have had, even more what sort of noises they would have made.

Interestingly, we *can* tell something about some dinosaurs, which one would think was completely unknowable: their social activity. If we look at the way in which dinosaurs behaved, from their remaining footprints and the way in which skeletons are found in groups, it becomes possible to suggest patterns of social interaction. It is, of course still only a suggestion, simply because we can never know, but just as Cuvier developed a comparative anatomy, we can suggest a comparative pattern of behaviour. So if the footprint patterns of dinosaurs were similar to a modern herd of, say, elephants, then it would seem reasonable that the herd moved in a similar manner to elephants.

The story of where the human race came from starts with a simple comparison of proteins that have been extracted from various bones. Here comes the leap: proteins only differ because their DNA code does. So why not look at the ultimate determinate of the organism, why not look at the DNA?

When looking at DNA it should be remembered that in very ancient material, nuclear DNA is often downgraded. Mitochondrial DNA often survives much longer. Once human mtDNA had been sequenced it would be reasonable to say that the variation could be exploited to determine relationships between different racial groups. This would be used to determine what the relationships between different modern racial groups are, and where different populations originated.

The starting point was, as is common in this sort of research, to get suitable material for analysis from any source available. Analysis of this material would hopefully show the range of variation within the human mitochondrial genome. The problem in analyses of this type is that although aboriginal populations around the world tend to be more insular, or at least they were, in Europe and the USA this is not the case. In the UK and the USA, and probably most of Western Europe, we are such a staggeringly interbred group, that virtually every conceivable variation can be found in these populations. With largely isolated populations the difference is such that although there will be variation, mutation rates are relatively low. Strictly speaking, mutation rates as such are not low, what *are* low are the mutation rates found in a *specific* area of the genome, whether nuclear or mitochondrial. So when looking at specific target sequences, the mutation rate is really quite low. There are hot spots of mutation within the genome, but these are few and far between.

What there is, however, is a molecular clock. The mutation rates are stable, so by looking at admittedly small areas of the genome, it becomes possible to see the way in which changes have occurred, but better still, the rate at which changes have occurred. So we can take the clock and generate a rough time scale for the divergence of different ethnic groups.

So now we are in the arena of, possibly, determining what our lineage is. Or at least, what the relationship is between different groups of humans. The original work was carried out using mitochondrial DNA, but this is fraught with difficulties. Using material from any ethnic group is going to lead to misunderstandings. This is a simple problem of logic. Mitochondria are inherited exclusively down the maternal line, so when mtDNA is tested the assumption is that the maternal line is continuous, but if it is not, the results will become equivocal. It is for this reason that mtDNA is just not suitable for analysis in a multiethnic society. There has been too much mixing to be able to say more than that you are what you want to be.

A consequence of this is that if there is a maternal quirk, in the form of a long-distant male relative having a child with a different mother, but being kept within the family, then descendants will have different mtDNA, so lineages will become confused.

An interesting aspect of the use of mtDNA is that it is possible to look into the past of Neanderthal man. The mutations in the mtDNA seemed far more common than would be expected in our own *Homo sapiens*. The inevitable conclusion of this was that these were not linear descendants. Current humans did not descend from Neanderthals, these were not missing links in the descent of man, rather, the Neanderthal man was a different species, from a far distant common ancestor. It would seem strange to us now to think of two quite different hominid species, occupying the same area, but having a different genetic make-up and hence characteristics. We do not know whether interbreeding was possible between these two species, what we do know is that one of them – Neanderthals – disappeared. This is evolution, and we have to recognize that they were not our progenitors. We both came from a common ancestor and are not related by lineage, but by diversification from this common ancestor.

Mitochondrial DNA can also show us relationships between different groups and populations, especially how island populations are related. It may even be possible to help clarify complicated questions of archaeology. One of these big questions of archaeology is the relationship between Pacific islanders. Hagelberg and Clegg, working in Oxford, addressed this question. Before looking at their results it would be worth looking in detail at the problem and the geography of the area.

The way in which the islands of the Pacific were colonized has long been controversial. The whole area was covered by a number of migrations, starting with the invasion of Australia and New Guinea and subsequently New Zealand. There are broadly two groups, which are genetically distinct. There are the Melanesians that live on the islands around Papua New Guinea and the Polynesians that live on the islands that extend deep into the Pacific Ocean. Such is the manner in which the various migrations took place that the Melanesians arrived many thousands of years before the Polynesians, so although they both originated from Southeast Asia, the Polynesians are genetically closer to Southeast Asia than the Melanesians are to either. This may reflect a different starting point for the two groups, or it may be genetic drift associated with a very small starting population.

What the Oxford team did was to use a characterized mtDNA mutation from Asia. This same mutation has been found in Polynesian samples taken from ancient skeletons, with the exception of very ancient skeletons from the central Pacific islands of Fiji, Tonga and Samoa. The mutation was not found in skeletons from Melanesia. It seems possible then that Polynesia was invaded, either totally or in part, by Melanesians, rather than directly from Southeast Asia. It would seem that in some areas the use of DNA does not always clarify a situation as well as we might like.

What this shows us is that while we might ask straightforward questions and hope for simple answers, the human race as a whole is never that simple, and usually neither are the answers themselves. We can only ever achieve a consensus on a hypothesis, which, since it refers to the past, can never actually be tested. That is not to say that we should not pursue these investigations. They serve a useful purpose in at least helping to clarify our ancient history. Part of the excitement of these studies is that archaeology and anthropology can come together tentatively to answer questions, and yet biology might (and often does) throw up some surprising results. We might find that what had been assumed to be single events, such as human migrations, or the rise of agriculture, become events which took place several times.

When looking so far into the past it is like looking at a beach. From a distance it looks uniform and simple, but as you look closer and closer at the sand it resolves itself into a more and more chaotic reality. There will never be an end to our curiosity about our past and so it should be.

7 TELLTALE DNA

When DNA profiling was first devised, one of the earliest uses to which it was put was to reconcile a long-standing problem that was elegantly stated by Henry Mayhew in 1851. This passage refers to 'river finders' or 'dredgers':

> The dredgers are the men who find almost all the bodies of persons drowned. If there be a reward offered for the recovery of a body, numbers of dredgers will at once endeavour to obtain it, while if there were no reward, there is at least the inquest money to be had – besides other chances. What these chances are may be inferred from the well-known fact, that no body recovered by a dredgerman ever happens to have any money about it when brought ashore.
>
> *London Labour and the London Poor* (1851)

When there was a reward offered then, the bodies would be recovered with the main aim being their identification. With no reward, of course, they would raid the bodies. The problem when recovering a body for a reward would be actually identifying it. This identification can often lead to a case being solved or a mystery being put to rest. The problem can be worse than that and much more difficult. If a body is dismembered, chopped up, or cut into pieces, how is it possible to decide how *many* bodies are present? To put it another way, three arms and a leg: is that one, two or three bodies? Part of the problem is that after a relatively short time in water bodies lose both their skin colour and later their fingerprints, as the skin lifts off. This results in what is essentially an unidentifiable body, using traditional methods of identification. It is often thought that a body weighted down in water will stay down, but without sufficient weight and depending on the water temperature the contents of the intestines can start to ferment, producing sufficient gas to cause the body to float. Eventually the tendency is for the body to sink as the less dense elements, such as skin and fat, decay and the overall density of the bones takes over.

It is not just water that can cause problems to those that are trying to identify bodies. Consider the case of a mid-air explosion, on a commercial airline. The disintegration of the aircraft both at altitude and speed would result in intense disruption and destruction. When a body drops from 30,000 feet the impact results in a crater. A body creating an impact crater will be barely distinguishable as human, so it would make identification of the individuals impossible without special techniques. It may be imagined that it would be possible to identify someone from dental records. Unfortunately there are several reasons why this is impossible. These involve such things as finding the dentist, finding the records and the probability that the head will probably be so badly damaged that dental records would not make any sense anyway. What is needed is something better. What is needed is DNA.

Along with the physical damage caused by high-speed disintegration there is also likely to be fire damage, just as there would be in a domestic or industrial fire. Fire is a very severe destroyer and eradicator of physical structure and appearance. So to be sure of identifying an individual after such disasters it is necessary to carry out family studies, so that identification can be revealed with some certainty. At least with airlines, accurate records are generally kept of who is on the flight, as long as the passengers are honest as to whom they are.

Of course, it is not just accidents that result in unidentifiable remains. Murder can also result in unidentifiable remains, especially if the body remains undiscovered for long periods, buried or not. There have even been cases of suicides by hanging that have been carried out in quiet and rarely visited places, such as a deserted and derelict warehouse. The body, inevitably, is not found for a considerable period and consequently rots while hanging. The process begins as with all bodies in the open, when flies arrive and lay their eggs on the corpse. With time, decay progresses and the body starts to disintegrate as the connective tissue gives way, giving the appearance of a stretched carcass. In such cases as these there might be a very good idea regarding who the victim or suicide is from missing person reports, or possibly documents which are still readable on the body. Failing additional information and since there is usually more than one person missing at any time, it is only by comparison with known family members of the suspected missing person that certainty can be achieved in identification.

When it comes to identification of remains of murder victims the range of techniques available are quite wide, but when it comes to decayed bodies it is necessary to turn to techniques developed for DNA analysis, usually used in analysing ancient DNA from archaeological remains. The first use of DNA analysis to identify skeletal remains took place in the UK in the late 1980s and early 1990s. The methods used to produce a DNA profile are in themselves interesting. The case involved the murder of a 15-year-old girl in 1981. Her body was buried, wrapped in a carpet. It was not until 1989 that her body was discovered, badly decayed, in fact so badly decayed that she was unrecognisable.

The first step was to try and establish a potential identification by facial reconstruction from the skull. This was followed by the use of dental records. This certainly indicated it was the girl that had been reported missing eight years earlier, but confirmation could only be made by using DNA to see if her DNA slotted into the DNA patterns which were found within the family.

A section of the femur of about 5gm was used in the analysis. The scientists working on the project were very aware that contamination was more than just a potential problem – it could ruin the entire test. With this is mind the surface of the bone was sandblasted to remove about 1-2mm of the bone. The remaining segment was divided into two so that all further testing could be done twice, in parallel. The DNA that was extracted was partly human in origin. The remaining DNA was presumed to be from the organisms that had caused the later stages of decay. These would be primarily bacteria and fungi since most if not all of the larger organisms such as flies and beetles would have dispersed long ago.

Once sufficient DNA had been produced for analysis, DNA from the mother and father was used for comparison. By knowing the profile of the parents it is easy to tell whether the body could belong to the family. The final result was that the DNA confirmed the other, earlier, suggestions that the body was the 15-year-old girl.

Identification of victims of murder has always been a problem when the bodies have been undiscovered for long periods, even more so when deliberate attempts to conceal a body, or bodies, are pursued. An event, or more correctly a series of events, took place in South Australia that shook the community and rattled the whole of Australian society. No longer was it thought that mass murder was something that didn't happen in rural Australia. The centre of the events took place around a small town called Snowtown. The way in which events unfolded was remarkable and the implications for the community both disturbing and deep.

It all started with the discovery of a body by a farmer in a field that he was ploughing. This in itself would not necessarily be so odd, in that like all inhabited ancient sites these things turn up from time to time. On this occasion, however, things were not so much archaeological as homicidal. The original body, and there were more, was found by the farmer in 1994. Being well decayed it was necessary for the pathologists to determine that the corpse was a young male. There was no known motive, reason or cause for his death, just as there was no known identity. This discovery took place at Lower Light, which is approximately 50 kilometres north of Adelaide, the capital of South Australia. The remains lay, unidentified, for many years, with the understanding that should it become possible to identify them in the future, material would be available for analysis. The body was eventually identified from X-ray photographs, rather than DNA profiling, in 1999. It turned out that the victim was Clinton Trezise, who had been reported as disappeared in 1994. His family has simply assumed that he had run away.

Worse was to follow on from this discovery and it became progressively more and more gruelling for all concerned in the investigation and the general public because this was going to involve some very small, rural communities, like Lower Light and Snowtown, with a population well below 1000. It was in May 1999 that, following a lengthy police investigation into a series of missing persons reports, the police visited the now disused Snowtown branch of the State Bank of South Australia. Located on the main street, it is a substantial brick building with its own vault. What the police discovered there appalled and astonished both them and, as word got around, both the town and all of Australia. Although closed, the bank had been rented and in the entrance area, where customers would have queued and conducted their business, there were computers and sundry bits of electrical equipment. It was not this material that caused such consternation. It was what was found in the bank vault that caused such amazement. Beside the stink when the vault was opened there was initially little to suggest what the contents of the six plastic barrels would be. They contained acid – and body parts. When the contents were looked at in more detail, it was decided that there were at least eight victims, some having been in the vessels for at least three months. Although the idea of using acid would have been to dissolve the corpses, the acid is slowly neutralized so that the barrels became almost like fermentation vessels, hence the revolting smell that was reported by the police.

Further from the banks, a rented house was also raided for evidence, but it was back in Adelaide that the next step in the grisly story was to unfold. It was only a day after the discovery of the bodies in the bank vault at Snowtown that the police visited the northern suburbs of Adelaide. Three different addresses were raided and three different individuals were arrested. Each of them was charged with murder of a person 'unknown' between 1 August 1993 and 20 May 1999. The three individuals on charges were John Bunting, Joe Wagner and Mark Haydon. They ranged from 27 to 40 years of age and were kept in custody with the full expectation that further charges would be laid, but there was one problem – the identity of the victims. It was going to be pivotal to the entire case, whether or not the dead were identified. It is surprising how persuasive it is for a court to know the name and origin of a victim, rather than the victim being referred to just as a corpse of unknown provenance. Later on, in June of that year, a fourth suspect was arrested. He was James Vlassakis, and was only 19 years old. He was held in custody in a secure unit of a psychiatric hospital. It was seen that once arrested he tried to commit suicide twice and so was deemed mentally unstable. Proceedings for all four of the defendants were scheduled to start in the early days of July 1999.

Later on the police were looking for more victims at the house that John Bunting had previously rented. The house has since been demolished, but at the time it was a scene of great activity. The building had an outside area that revealed a pit. This pit first revealed a body at about two metres deep, but then went on to reveal a further body a metre below the original find. This was the result of a long, covert operation, originally to investigate the disappearance of three people: what was emerging was rather more than that. As the investigation proceeded, so too did the range of police groups that become involved in the developing investigation. There was at this point a known group of ten individuals that had been murdered, although their identities were not necessarily known. Some were identifiable by the more traditional process of fingerprint analysis, even though the skin and the fingers must have been badly degraded. That fingerprint comparisons were possible infers that the victims either had a police record or there was another bank of personal identification requiring fingerprints available to the police.

Being a group of missing persons it is possible that finger-prints were available for just such an eventuality.

The original three missing persons that started the whole investigation were Barry Lane, a 40-year-old trans-vestite and paedophile; Clinton Trezise, who was last seen at the age of 22; and Elizabeth Haydon, 37, a mother of eight. That Elizabeth Haydon shared a surname with Mark Haydon is not a coincidence; she was in fact Mark's wife, and had been reported as missing in 1999. As can be imagined, South Australia was horrified with the gradu-ally increasing body count, with ten bodies already found. The feeling amongst the general public was that there might be no end to the hideous story unfolding day by day. It had become necessary to identify the bodies with DNA methods. Bunting and Wagner were sentenced to multiple life terms in September 2003. The jury failed to reach a verdict on Haydon, so he was remanded in custody and, as of May 2005, was still awaiting a retrial.

In the case of an aeroplane crash, found bodies can be identified by a process of elimination, assuming there are known family members who are prepared to donate samples so that a family tree can be constructed from the DNA profiles. Using this information it becomes possible to say whether one of the bodies fits into the constructed family tree. With a number of known bodies and a number of unknown bodies, we need to simply fit the two together and the identity of the body is revealed.

A forensic psychologist was asked for his opinion regarding the motive for the killing. His opinion was that since there were several people involved in these murders, it was unlikely to be a simple case of a murder with a sexual motivation. However, it might well have been a part of the motivation and it was assumed by the public that it was, as the background of the accused became more widely known. What the forensic scientist suggested was that when three or more offenders were involved the motivation was not so likely as to be sexual as financial.

The use of DNA to identify the bodies was progressing well, such that all of the bodies from the Snowtown bank vault had been identified, but the murders still hung over the town. They did not seem to be random killings. The victims seemed to be known to the four accused. Continuing collection of Social Security cheques of many of the murdered victims attest to the assumption that financial gain was at the bottom of the process. It is often quoted in psychological textbooks that once the first act has been carried out the second get easier. Once the Rubicon has been passed and several killings have taken place, there seems no point at being overly squeamish at the next one. If they hadn't been identified, with DNA methods proving a powerful tool, the accused murderers might have had no case put against them.

It is still not entirely clear that money was the complete motivation. *The Daily Telegraph* reported that some of the victims had gags stuffed into their mouths and rope around their necks. As well as this there were burns on some of the bodies and hands and feet had been cut off. There was also equipment discovered that was capable of giving a severe electrical shock. So although the bodies had been identified by linking the DNA profiles from the victims to families via the construction of a family tree for each of them based on a missing persons report, it will probably never be known what the motivation of the murderers was. Neither would it become known how they came together to carry out these crimes or how they killed their victims.

In 1987 in the USA a very early use of DNA as a method of convicting a criminal involved a double murder and a rape. It all started when Randall Jones found that his car was stuck in mud, stuck so firmly that it was only going to be moved if he could get a tow. Nearby there was a parked pickup truck. Jones approached the vehicle and found that inside there was a man and women, reportedly in their early twenties and asleep. Jones shot them both through the head at close range with a rifle. He then dragged the bodies into the woods and towed his car out of the mud with the truck. Approximately 40 minutes later Jones returned to the bodies and 'raped' the dead woman. It was the semen left after the 'rape' that was used to generate a DNA profile of the criminal. At the time there were three laboratories in the USA that carried out DNA analysis. These were Cellmark, Lifecodes Corporation and Forensic Science Associates. Cellmark in this case carried out the work. Although such work was virtually unheard of outside scientific circles, the jury seemed quite happy with explanations of the science and took only 12 minutes to return a verdict of guilty. There was further evidence in this case however. Sometimes, even where DNA evidence points a finger, it can be difficult to convict an individual on that evidence alone. There was a case where an individual disappeared and the only clue was a bloodstain found in another man's house. In this case, with no body, the claim was made that he could still be alive, so the accused was acquitted. It was much later that the skeleton of the disappeared man was discovered, but too late for legal action.

There is another aspect to DNA evidence that is at least as important as finding the guilty. That is exonerating the innocent. It is not unusual for someone arrested for a crime based on eyewitness accounts and other such evidence to be excluded from the investigation once a DNA profile has been produced from the scene of the crime and the arrested individual. This leads to a disturbing thought. If DNA evidence had not been pursued it was likely that there would have been a number of people convicted on evidence, even by their own admissions of guilt, who were in fact innocent. It was for these and other reasons that the Innocence Project was set up in America. The aim was to make sure that innocent individuals were not incarcerated on spurious evidence, including confession. Making a confession for something that you have not done may seem rather odd, but clever verbal footwork on the part of interrogators can undermine an individual to the point of confessing just so that it stops. The Innocence Project has made some remarkable changes to many people's lives by using DNA evidence to demonstrate, sometimes many years after the crime was committed, that the convicted person was innocent after all. Such work as this is extremely valuable for both the individual and society since it is not fair to lock up someone who is innocent and it reflects very badly upon society to lock up people on tenuous evidence. But at least in the USA steps are being taken to rectify injustice. It should be added that the Innocence Project was started not by the state but by two private individuals that were also lawyers.

In Canada the problem is tackled in a slightly different way. If it is demonstrated that an individual has been wrongly convicted they will go back over the case to find out why the wrongful conviction occurred, with the stated aim of trying to make sure miscarriages of justice do not occur in similar circumstances.

One thing that that DNA analysis cannot easily tell you is how long a body has been dead. It has been tried to tie together the state of breakdown of DNA and how long a person has been dead, but it is extremely difficult because it depends so much on the way in which a body is dumped or buried and the environmental conditions in which the body is dumped. There is a method that has been used for many years to give an idea of the age of a partially decayed body, although once a body is so far decayed that it is little more than a skeleton, the dating process breaks down.

This technique is forensic entomology. It was first used in the UK in 1935. On 29 September 1935 two bodies were discovered in Dumfriesshire in Scotland. At that time there were no sophisticated biochemical tests that could be used to determine the time of death, but there was a clue. The bodies were infested with maggots, which pointed to the bodies having been dumped in a ravine between 12 and 14 days previously. The bodies were identified as the wife and maid of Dr Buck Ruxton. Other evidence eventually led to the arrest and conviction of Ruxton for the murders, for which he was hanged.

Of course it is not quite as straightforward as simply looking at the maggots present on a body and making a pronouncement. As flies do not regulate their own temperature the rate of development is directly related to the ambient temperature. Consequently, the forensic entomologist needs to know the temperature cycle, both day and night, that had been present since the body was dumped. For example, below 4°C development and growth of maggots virtually stop. There are also other things that the presence of maggots can indicate. Blowflies have delicate eggs, so they are laid in preferred places, such as eyes, nose, mouth, ears and anus and frequently underneath the body as well. Stab wounds and gunshot wounds are also favoured sites for egg laying. The presence of maggots in unexpected places may indicate a cause of death because the very action of the maggots will eventually destroy cut marks and gunshot wounds.

Much of what we know about decomposition of bodies, by flies, fungi and bacteria, comes from the USA. I do not imagine that most people leaving their bodies to science for research expect it to be left in a field to rot, but actually for a surprising and increasing number of people, this is precisely what they want to happen to their bodies, to do something useful after death. It is important to know how such processes as decay progress, so that when a murder victim, or a dead vagrant, is discovered it is possible, with some degree of certainty, to know how long they have been there.

This information has been gleaned from an institution called 'the Body Farm'. This is not its real name: officially it is the University of Tennessee Forensic Anthropology Facility, but ever since Patricia Cornwell described it in her book called *The Body Farm*, that is what it has been known as. The facility is three acres at Knoxville, Tennessee, surrounded by a high wooden fence and razor wire, not far from the University of Tennessee Medical Centre. It was started late in 1971 by forensic anthropologist William Bass. Although Bass retired in 2000, he still stays active, lecturing pupils and still takes an active interest in 'the Body Farm'.

At any given time there are usually about 40 bodies within the area of the body farm, all there to try and determine not only exactly what happens to a corpse as it decays, but also what the time scale of events are. The bodies are buried in shallow graves, submerged in ponds, stuffed into car trunks, left in the sun or left in the shade and even just covered over with undergrowth. They are not left unattended of course, as careful and meticulous measurements are made throughout the process of decomposition. Students and staff, anthropologists from the University of Tennessee, carry out measurements. The measurements range from when insects of different species arrive, not just flies but beetles and all the predators of the scavenging decomposers themselves. Besides this there is considerable biochemical data collected, such as the rate of protein degradation as well as degradation of organs and gas production from putrefaction. It is reportedly a very, very smelly job, but strangely popular amongst the students.

Interestingly, some states in the USA are large and sparsely populated, so when human remains are found they tend to be skeletons. In other states, such as Tennessee, that are smaller and more densely populated, bodies are often discovered long before decay has completed, so it should be possible to produce an educated idea of how long the body had been left. One of the findings is that there can be two different rates of decay: internal putrefaction from the digestive enzymes of the body and insect-mediated decay, mainly by fly larvae. Although during the summer months there always seem to be flies about, if a body is in a house with little or no access for flies, or one of the solid hanging insecticide blocks, it may be some time after internal putrefaction has started that the flies can get in and start work. By using the knowledge gained from this work it has become possible to be specific about time of death and this has helped many law enforcement agents determine who was at the scene when the deed was done.

A most unfortunate, but very necessary, use of DNA as an aid to identification came with the regrettable case of a corrupt crematorium in the USA. The crematorium was called the Tri-State Crematory, and the unfortunate family were the Hardens. One member of this extended farming family was Lloyd Harden. To give an idea of the extent of the family, Lloyd was one of nine siblings, both male and female. Part of the story starts when Lloyd was accidentally shot by an elder brother when he was in his teens. The bullet was never removed and so stayed with him until his early death at the age of 44. The family wanted Lloyd to be cremated and so approached the Tri-State Crematory to carry out the task and return the ashes to the family. It had a wide catchment area because it was more or less at the confluence of the three states of Georgia, Alabama and Tennessee, at a town called Noble. The cremation was duly carried out and the ashes returned to the Harden family.

Nearly two years after this, with the ashes of Lloyd still kept by the family, a story began to unfold early in 2002, involving the Tri-State facility. A deliveryman had been to the crematorium and reported that he had seen a human skull in the grounds. The first point of contact was the Environmental Protection Agency, who quickly realized that this was a major problem and required a great deal more investigating than they could carry out. It became a police matter. As the story unfolded, it became a sorry tale of greed and deception. In all more than 300 bodies were discovered within the grounds of Tri-State Crematory. The bodies were all over the place, effectively dumped in the most disrespectful way. It is worth noting that they were not the victims of murder or anything of that sort. They were the dead victims of a crematorium owner who had not been honest enough to complete the contract of cremation and had just dumped the bodies. It was for this reason that the first charge bought against the owner was one of obtaining money by deception.

It was only a few days after the start that the Governor of Georgia declared a state of emergency in the county. During this time there was considerable media interest. Hundreds of relatives who thought their deceased family members had been cremated there and who held the ashes now faced the realization that they did not actually hold the ashes at all. The main question was, of course, 'if these aren't the ashes of my relative in this urn, who or what is in here?' Of the bodies found, just under a quarter were identifiable relatively easily, because they had not been dead that long. For others it was going to be a far more difficult task. Not long after these discoveries, a civil legal action was started. At the same time the relatives of Lloyd Harden were becoming increasingly concerned regarding the true final resting place of their relative.

The first task was to find out whether the ashes were actually human. After all, there were a lot of bodies and a lot of people with remains they genuinely thought were the cremated remains of those bodies, but actually were not. When the remains were analysed, it turned out that what was in the urn was as far as could be determined not human. The major component of the material seemed to be cement powder, which has the same grey, dusty consistency of cremated remains. There was some material added as a filling agent, too.

Given that the material in the urn was not the remains of Lloyd Harden, it seemed reasonable to assume he was among the bodies recovered from the grounds of Tri-State Crematory. After two years in the open, decay would have reduced the body to a state that was unrecognizable. With this in mind, the investigating authorities had taken samples from every body that was discovered. These were submitted for DNA profiling. To discover from these profiles the identity of the bodies it was necessary to take samples from volunteers of members of the families with a known relationship to the deceased.

With all this information, the next step is to construct a family tree, in the same manner as if we were conducting a paternity test. All of the parts of the profile found in the body should be explainable by inheritance from the two parents. Even if the parents are not available, a genetic family tree can still be constructed by using samples from brothers and sisters, nephews, nieces, uncles and aunts. Although not as clear-cut as using material from parents to produce a DNA profile for comparison, in the case of the bodies at Tri-State this uncertainty was not a significant aspect of the analysis. The reason: if the DNA profile from a single body matched the profile which would be expected from the profiles generated from family members it would be safe to assume that the body was a part of that family. The documentation was sufficiently clear to know who had been sent there for cremation. It was a simple case of matching the bodies to the families.

With Lloyd Harden, his early death had resulted in an autopsy being carried out, and the samples from this had been kept. Marrow was used from his body to create a profile, and it was matched up with the profile created from the autopsy samples. Lloyd Harden was cremated properly two years after his death.

There is a sad story regarding the bringing to justice of a rapist and murderer. While all such events are unpleasant, this was particularly so because of the long sequence of events, perpetrated by an individual that can only be described as vicious and remorseless. It also took decades to bring him to justice. There is also some irony in the story because the victim worked with DNA. She was a gifted scientist who saw the potential of DNA for the benefit of mankind.

The events took place in California over a relatively long period of time, but they started on 7 April 1984 when Dr Helena Greenwood was sexually assaulted in her own home, being forced at gunpoint to commit oral copulation with an intruder. Greenwood was not an American; she was from England, a molecular biologist, brought up and educated in the UK and at this time working in industry in the USA. This was an unprovoked attack by an unknown assailant. Little happened for quite a while after this assault for Helena and her husband. Helena moved job to work with a different company. Her move, so far as we can know, was not motivated by the assault, although it would seem reasonable that she would not want to stay in the home where she had been so brutalized. Her new job was still in California and with her new position and new home she put the past behind her. During the break-in before the assault, the attacker had come in through the kitchen window and had handled a teapot. This had left a fingerprint on it. After the attack, this was the only tangible clue other than a semen stain left by the attacker on the bedclothes. Although at this time no DNA analysis was possible, there was some information that could be gained from the semen.

In the meantime there had been a number of incidents reported, broadly of the same type, but without the humiliating ending that had overtaken Helena. The accused in these cases was Paul Frediani and he had finally been arrested for exposing himself to a young girl. When his fingerprints were compared with those found on the teapot at the home of the Greenwoods, they matched. So just after over a year, on 9 April 1985, Frediani was arrested for the sexual assault on Helena Greenwood and bail was set at $100,000.

The preliminary hearing of the case started on 7 May with Helena giving her testimony. It was the next day that Frediani was brought in to the court. It should not be underestimated how distressing it is for the victim of these sorts of assaults to appear in court. So very often it is described by the victim that the cross examination is as distressing as the event itself.

During this aspect of the prosecution case it was revealed that analysis of the semen found in the bedroom revealed a blood group type of O secretor. The secretor gene is a dominant gene that results in the secretion of blood group types in body fluids such as saliva and semen. O secretors are relatively common, so more enzyme analysis was required. After the hearing the presiding judge at the hearing was persuaded to reduce the bail from $100,000 to $25,000. This was a sum that allowed Frediani back out into society.

A pre-trial conference was scheduled for 4 September to deal with the sexual assault charge, but before that, on 22 August 1985, disaster struck Helena Greenwood again. While in her garden, Helena was strangled to death. This was made all the worse a situation because it was her husband who discovered her body. Having put the sexual assault behind them the terror and horror had returned, but now Helena was no longer there: it was her husband alone. The usual detailed forensic and police investigations were undertaken, which turned up some fragments of fingernails. These were duly found, collected and bagged. What was apparently obvious to the police was that there had been quite a struggle before Helena was strangled; she had put up a fight, a fight for her life that she ultimately lost. The customary autopsy of the victim later confirmed the first impression of the police that it was murder: death by strangulation.

After all the necessary documentation and administration was carried out Helena Greenwood was cremated. Her ashes were taken back to England by her father and interred with her mother, who herself had only died a few weeks earlier.

This put the prosecution into a very difficult position. They were determined to find the truth of the accusations. After all, one of the most important aspects of any state is the defence of its population, and they had failed Helena. It would be normal for the victim of a sexual assault case to appear in person. In this case of course, this was not going to be possible. So when the trial started on 14 October 1985, what was possible was for the *verbatim* statement Helena Greenwood had given at the preliminary hearing to be read out by a member of the court staff. This is what happened, without the emotive knowledge of the death of Helena being given to the jury. The various examinations of witnesses went on until the summing up by both the prosecution and defence, followed by the judge instructing the jury in their duty.

In various states of various countries, the jury has been dispensed with, for example India. This does not mean that justice is not done, but there is a feeling in English law that a group of peers acting as a jury offers a more balanced judgement. In places with as many languages as India, with in excess of 1,600 different languages and dialects, of which 18 are officially recognized, held together by a single democratic ideal, any jury picked at random would quite likely have individuals in it that did not understand what was said. There is a way around this that seems to work very well – do not have a jury and use a language in court that everyone agrees upon.

In the case of India, the language is English. Incredibly, more people than there are in England speak Indian English. With the Indian court system, as long as the aim is the same, to find the truth, and as long as politics are excluded from the judiciary, then hopefully the result will be fair, honest and just. In the Frediani case, the jury was out for more than nine hours. The result was that Frediani was found guilty of forced oral copulation and the use of a firearm during a burglary. Bail was set at an increased rate of $75,000 until the sentencing hearing.

It was three weeks later that Frediani was back in court to be sentenced. As could be imagined the prosecution wanted a greater sentence than the defence. The final result was a sentence of nine years in prison. But the story does not end there, as there is always the possibility of an appeal against a sentence. By raising the bail, Frediani was released from prison pending his appeal. It was on 16 January 1987 that the appeal court made its pronouncement. This was that reversal of the judgement should be made, on the basis of legal procedures that the police had committed violations with regards to the manner in which the interview of Frediani had been carried out. In many ways this was just a simple technical detail, but one that proved very costly.

One of the rules in these situations is that a mis-trial can be re-tried and so it was that on 29 July 1987 the second trial started. Like all these proceedings, courts take their time: it essential that all parties in the case have their say, how ever much we might not like the idea. By 5 August 1987 the case was finished, the various lawyers had completed their summing up and the judge had instructed this new jury in their responsibilities. This time it took only four hours for the jury to find Frediani guilty, on exactly the same charges as at the first trial. Once again the case went to the California appeal court and once again the conviction was overturned. There was now a very real possibility of yet another trial. This was early 1989 and by legal plea bargaining by Frediani and his lawyers another trial was avoided, which also meant that Frediani would be out of prison by July of the same year, 1989.

This process of trial and appeal, backwards and forwards, had taken so long that by 1988 a law had been enacted that required all sexual offenders to give a blood sample to the state for either analysis or storage. It would not usually be analysed but kept on file for future reference. This was two years after the first use of DNA analysis in a criminal case and the science had already been wholeheartedly embraced as a powerful tool.

Like many law enforcement agencies in many different countries, an unsolved murder is never closed. Of course, some murders will never be solved and become lost in the mists of time, like those of Jack the Ripper. Identifying who Jack the Ripper was is still something that occupies column and book space, but the sheer lack of direct evidence that prolongs the debate also means that the issue will never be closed. Sometimes what seems like an insuperable problem can be levered open by a new and radical technique, which is what happened in the case of the murder of Helena Greenwood. Ten years later the murder of Helena Greenwood was being looked at again with a view to using DNA profiling, and so it was in 1999 that the retained fingernail clippings of Helena and some of her hair, complete with roots, were sent to a testing laboratory for analysis. It was important that the hair had intact roots as this is the only part of the hair that contains nuclear DNA, which is what is needed for producing an STR profile. This would produce a DNA profile that could be subtracted from any mixed profile that might be generated from DNA found associated with the fingernail clippings. Much to the delight of everyone involved in the case there was DNA found on the clippings, which did not belong to Helena and so it was assumed that this had originated from her attacker as she fought for her life. Just as material from Helena had been stored in a deep freeze for nearly fifteen years, so, too, had the blood sample from Frediani been kept, for ten years, deep-frozen after his final release. The blood was sent to the American laboratory of Cellmark Diagnostics. The analysis of the blood sample was carried out at a different laboratory to the fingernail clippings, but being carried out using STR technology meant that the results from the two different laboratories could be directly compared. By early December 1999 a convincing result from the analysing laboratories had been reported: a match between the DNA found on the fingernail clippings and the blood sample taken from Frediani ten years earlier. An arrest warrant was issued and the police department swung into action. Frediani was arrested for murder.

It was 15 March 2000 when Paul Frediani was taken to court for the preliminary hearing. The details recounted to the judge mainly revolved around DNA, genetics and the analysis carried out by the testing laboratories. Unlike the practice in the UK, the American authorities took the calculated probability into court. In the Frediani case, the probability was 1 in 2,300 million million. The reason that this is not done in the UK is that numbers of this magnitude are virtually impossible to understand. Most people operate on the basis of numbers up to a perceptible, human, figure and then understanding of the number is lost and it just becomes 'very big'. The more human practice is for the forensic scientist reporting the odds to say that the results give no support, moderate support, strong support or very strong support for the proposition that the DNA came from a specified individual.

After a long wait the trial started on 11 January 2001 with the appointment of jurors, although the trial proper did not get underway until 16 January. After two weeks the trial came to an end and with the judge instructing the jury in its collective responsibilities. The jury left the courtroom to start deliberating upon the evidence that had been presented to them. It was during the middle of the afternoon that the jury returned to the courtroom having made a decision upon the verdict. This was read out by the foreman of the jury: guilty of murder in the first degree. It was a unanimous vote, so a date was set for sentencing: 19 March 2001. When the time came Frediani was sentenced to life with no parole. This unfortunate tale has other unfortunate twists. Helena's husband died of cancer before her killer was found and her father, also suffering with cancer, held onto life until after the verdict, but died but a few hours after hearing it.

One aspect to STR analysis that is rather unsettling is when an exclusion is ignored. Such a case took place at Pudukkottai in the southern state of Tamil Nadu in India. Pudukkottai is an hour's drive from the nearest large town, Turichirapalli, known locally as Trichy. This is a six-hour train journey from Chennai, previously known as Madras.

The case, in brief outline, revolved around allegations of the rape of 13 young girls and the murder of a Sri Lankan by the leader of an Ashram, Swami Premananda. The consequence of one of the alleged rapes was a pregnancy, which was terminated. This resulted in arrest and charges being filed against Premananda in 1994. Events unfolded in a large courthouse, open on two sides, with an ancient ceiling fan gently stirring the hot and humid air. That there was such a large court building in such a relatively small town is for purely historical reasons, dating from when Pudukkottai was a city state. These sorts of court proceedings are carried out in English, as mentioned before. A legacy of British rule has resulted in the wearing of gowns by the lawyers, but they do not wear wigs, which would be far too hot.

Various legal arguments in the case were put forward

until it came to the point when it became obvious that it was going to be necessary to carry out a paternity test to determine whether Premananda was actually the father of the aborted foetus. This initial analysis was carried out in a laboratory of the University of Hyderabad in 1996 using a number of untried probes to produce single locus and multilocus DNA profiles. These, sadly, gave a very inferior result that could not be accurately interpreted, although the scientist reporting the results insisted that he could demonstrate a match. Beside this analysis there was also an STR analysis carried out in the same laboratory, unfortunately only using a single STR, which again it was insisted showed a match, indicating that Premananda was the father. At this point the senior barrister, Mr Ram Jethmalani, who was also an MP in the Indian parliament, decided to instruct an independent scientific expert to look into the DNA evidence. The conclusions were damning: the profiling had not been carried out in a forensic laboratory, but by a technician who was not a forensic scientist, using probes that had not been validated for forensic use and STR analysis which was completely unproven to have any probity of any sort.

By pleading to the court it was agreed that the defence expert could take material back to the UK for independent testing. This included a blood sample from Premananda and samples from the frozen foetus. Extraordinarily the blood sample was taken in court, so that everyone could be sure the blood was taken from the right person. The samples were sealed and taken back to the UK where testing in an accredited laboratory could be carried out. Being highly trained scientists it was obvious when the foetal material was looked at it was quite likely that we would be able to produce two profiles from it. This would be one from the foetal material and one from the liquid blood, the second of these most likely coming from the mother. When the six-panel STR analysis was carried out in London on what was now three samples, it was quite plain that we had three clear profiles. By taking out the mother's contribution to the profile of the foetus it was obvious that Premananda could not be the father. A report to this effect was written and presented to the court. It was necessary to return to court for cross-examination of the results. These were in direct contradiction of the conclusions of the Indian scientist. This was not a disagreement on a small point of interpretation but a massive gulf. The differences were a result of the difference in the way the analysis was carried out. We had used validated techniques in a Home Office approved forensic laboratory.

The atmosphere became a little heated to say the least and when it was obvious that the UK results had been carried out in the most stringent manner, the prosecution counsel got a bit personal with an outburst.

'I put it to you Dr Wall, you are not even a proper scientist.'

To the credit of the judge, she did reprimand the lawyer. At one point towards the end of the hearing the girl whose foetus had been tested came to court as a potential witness where she said that she would not give evidence. This was understandable as the court was full of spectators. These were of two broad groups, those that had heard about the case locally and were simply curious and those that had heard that Jethmalani was appearing. The second group were mostly gowned lawyers who knew of Jethmalani by his reputation and wanted to see him in action. Although not prepared to appear as a witness the judge asked her a single question: 'was Premananda the father?', to which she replied 'no'. The arguments ran on until the final summing up by the lawyers of both sides. The judge made the final decision as to guilt or innocence alone. The result was a disappointment: he was declared guilty as charged and sentenced to twenty years in prison. Even Jethmalani was incensed by the result. In the written judgement it was said that there seemed no reason for an Indian court to accept the findings of a foreign scientist over an Indian one. At the time of writing, Premananda is still in prison, pending an appeal in the High Court in Delhi.

Another case, which has caused some dispute regarding DNA evidence, is the rather higher profile case of O.J. Simpson. This is a relatively well-known case, but the details may not be entirely familiar. It was probably the single most significant collapse of a role model in recent times.

Orethal James Simpson was born in 1947 in San Francisco. He played American football at the University of Southern California. It was while there that his speed was recognized as a significant football skill. In 1968 he received the Heisman Trophy, being judged the best college American football player. In 1969 his career was in the ascendancy when the Buffalo Bulls, a team in the National Football League, managed to obtain his skills on the National Draft. To briefly explain the term, this is where the most promising players are sent in a democratic manner to major league clubs, to avoid a monopoly. During his career, Simpson shone, with records of all sorts during the 1970s. Supposedly due to frustrations in the inability of his team to make it to the finals, Simpson requested a transfer from his current team to a team on the West Coast. The club was the San Francisco 49ers, which he joined in 1978. At after two seasons he retired at the end of 1979, due to injuries.

Now to understand the fall, it has also to be understood

that this man was a towering role model for black youths, success in sport being following in later life by success as a media pundit and actor. He later appeared in such films as *Towering Inferno* (1974), *The Cassandra Crossing* (1977) and a series of comedy films under *The Naked Gun* house title. This apparently inexorable rise in fortunes was about to fall apart. On 13 June 1994, the Los Angeles police made an alarming discovery. At the home of O.J. Simpson's estranged wife were two bodies: one was Nicole Simpson, the estranged wife, and the other was an acquaintance of hers, Ronald Goldman. Both of them had been stabbed to death. O.J. Simpson was subsequently arrested and charged, but not before what can only be described as an ill-advised attempt at escape in his car, and the subsequent car chase, which found its way on to live national television.

One of the major players in the defence of Simpson was Peter Neufeld. He was a lawyer, who in 1991 helped set up the Innocence Project, which originated when he was at law school in New York. This was a network whose aim was to demonstrate the innocence of convicted individuals using DNA evidence. Although there was a perfectly sound DNA profile from various bloodstains found in the Simpson case, they were criticized, quite correctly. DNA profiling is a highly skilled activity and it requires extreme sensitivity. It is, without doubt, of immense probative value, but only when carried out properly. This does not start just in the laboratory, but when the original samples are taken. It is simply not right that a sample should be picked up, handed around and then analysed, with the result still being accepted as a true and faithful one. The Los Angeles police department admitted to mishandling all of the blood evidence. It was simply not possible to have the necessary level of confidence in the courtroom results that would be required. This apparent acknowledgment by the Los Angeles police department has had one result – all law enforcement bureaux in the USA have tightened up their acts.

Although there was a considerable amount of evidence in the case, the pivotal nature of the questionable DNA results seems to have swayed the jury to such an extent that Simpson was cleared of the criminal charges in October 1995, with them having started in January of that year.

Strangely, in 1997, Simpson was found liable for the deaths of Nicole Simpson and Ronald Goldman in a civil court case, with the result that he was ordered to pay the victims' families the staggering sum of $33,500,000 in damages. DNA can both help and hinder.

A case where the use of DNA to track down a culprit can be found is in the unlikely situation of a rising rock star being both raped and murdered. Mia Zapata was 27 years old and a very popular singer locally on the US punk rock circuit. She was the lead singer with the band 'The Gits'. The events took place on 7 July 1993, when Mia was raped and brutally beaten to death in a quiet Seattle street. At the time of the assault there did not seem to be any specific evidence that might point to a particular individual. However, there was a twist to this investigation. The local rock community produced $70,000, which was going to be used to hire a private investigator, who would try to track down the killer of the singer. The money that was raised in all manner of different ways and in all manner of different amounts was used to pursue the perpetrator of the crime through a non-profit organization. It was even aired on the American television programme *America's Most Wanted*. All of the attempts to find the killer failed. The original idea was that Mia knew the killer, maybe through the music scene, but this turned out to be incorrect. The murderer was a 48-year-old fisherman from Florida, of the name Jesus Mezquia. He has a history of sexual assaults and presumably imagined that the rape and murder that he had committed in the far northwest corner of the USA would not be traced to the far southeast. He was wrong.

Once identified as a suspect it became clear that Mezquia has been in Seattle at the time of the assault. In fact he was even linked to the area where the assault had taken place. The DNA analysis had been carried out on saliva swabs taken from the body and preserved for many years. Even though at the time saliva was not regarded as a suitable material for analysis, the swabs were kept. At the time the lack of semen from the body made DNA profiling an almost impossible task. It was only with the introduction of newer and more sophisticated techniques that the saliva could be profiled. Once profiled the police could approach the National DNA index, the equivalent of the UK national database, to see if there were any matches. There were no matches at that time, but things were going to change quite quickly.

During September 1999, not long after the setting up of the database, a probation officer in Florida requested a sample for DNA testing from one of his clients – Jesus Mezquia. This was profiled, put into the DNA database and found to match the saliva samples from the dead woman, which had, by that time, been profiled. A week later a match was found between the saliva left on the body and Mezquia. It was now necessary to fill in the gap of three thousand miles between Seattle and Florida. It did turn out that Mezquia had been there at the right time and also that he had a history of vicious assaults. These included an assault in 1986 and an attack on a spouse in 1989, an assault to commit rape in 1990 and an attack on a woman who was seven months pregnant in 1997.

DNA is impressive, but the more times that we use it, the closer we come to that instance when we get the wrong person. What we have to take with us, both experts and the general public, is the central tenets of DNA analysis that the preceding case studies have set down. With DNA testing, we can never say for certain that it was definitely this person or that person. We can give odds on the likelihood of it being them, and these odds are often in excess of a million to one, enough (hopefully) to sway a fair and broad-minded jury. We can also offer total exclusions, saying in effect that a certain person is not linked with a crime scene of a sample from that crime scene. In between, as ever, is a grey area that must be trodden carefully. When odds dip too low, and mistakes are likely to happen, it is time for DNA analysis to step back in importance and allow other pieces of evidence and other techniques of analysis to hold sway. The humble fingerprint can still offer damning proof. The blood test is still available, and there is of course the evidence that solid and firm police work can provide.

DNA is a double-edged sword – most tools of this nature and of this power are. Where it can help, it can also hinder, and where it can offer a look back into the past, it can also offer false hopes. What the next chapter attempts to explore is the ethical issues that DNA is throwing up, including cloning, DNA Identification cards, and the dangers posed when lay firms get hold of scientifically complicated DNA and genetic data. What this (long) chapter has hopefully highlighted once and for all, is that DNA is a natural detective – it can hunt and seek out genealogies, family trees, evolutionary patterns and criminals in one fell swoop. Scientists working with DNA are DNA detectives by virtue of their very field.

8 WHAT THE FUTURE HOLDS

There are many ethical questions associated with DNA and DNA analysis. During a case study, it is not the best time to attempt to answer these questions, only to point them out. However, it is both worthwhile and important to look at, and to attempt to answer, some of these questions and I am sure that there will be many more that you can think of. What the book will focus on here is a particular aspect: the relationship between state and individual, in terms of DNA. In more general terms, the issues that the growing presence of DNA analysis in our lives raises will be addressed. It is the state that takes the major role in using DNA as a means of detection of felons. They also govern DNA legislation. Caution needs to be exercised by all individuals when a state, any state, claims dominion over such a powerful technology.

One of the problems that philosophers have when it comes to questions of ethics is that they have no specific answers. To more modest thinkers, like myself, ethical questions often have to be attached to necessary answers, and necessary compromises, rather than the broad-brush theorizing that traditional philosophy can afford to indulge in. The answers that are given here will be answers in keeping with my scope, answers that hopefully we can apply in our everyday life.

With ethics as a part of philosophy, what we have is a

system based on three broad questions, which to most of us have no clear answers. The first is the moral question, 'should I do that?' The second is getting a little less clear, 'what should I think about someone else's idea?' The third question that is traditionally asked is 'what do the words "right" and "wrong" really mean?' The ideas that have been proffered in order to answer these concepts are complicated, but we can all understand them and we are all allowed to have opinions about it. What we need to do is bear the questions with us.

With the powers that be laying claim to the technologies and advancements that genetics brings, we must address the following question: can we, at any point, be assured of the future? Well, we can try, but it is neither easy to think it, nor possible to be sure of it. Whenever a state, company or person says that they can insure you against the vagaries of the future, people should beware. If you think back over our history, that is the history of the whole of humanity, not just our twentieth and twenty-first century, then we have plenty of counter-examples.

Consider what has taken place in the past. If we take the lessons of the past and project them to the future we can say only one thing that is certain, that we do not know what will happen in the future. That we cannot guess what is likely to happen in any political state is a very good reason, for example, for not having compulsory identity cards. This was suggested in Great Britain during the early years of the twentieth century, but from what the politicians have said it would rapidly develop from a simple photo-card to something rather more all encompassing, say with biological information such as a DNA profile and other information which can unequivocally identify individuals. Such an identity card is said to be the ideal way of combating crime. I can never tell whether this is said ironically or not: if identity cards had any effect on crime it would be safe to assume that countries that have obligatory ID cards would have a lower crime rate, which they do not, neither do they have a better 'clear up' rate of the crimes that are committed.

Such information could be used in a way to persecute a specific group of individuals, or even to perpetrate genocide. It may sound like a glib turn of phrase, but genocide has happened more times in our recent history alone that most people realize. To allow people access to details of ancestry – actual genetic ancestry – is to provide a valuable and dangerous commodity. To start down this track when we cannot know the future nature of society we may be causing such problems for future generations, such that even the most enthusiastic backers who currently believe ID cards are a good idea would end up sincerely apologetic.

When we look at the use of DNA today, we can see several potential problems for the future. Primarily the major problems revolve around the assumption that the current state is a benign state and will stay that way. We can see from the past that this is an assumption upon which we cannot depend. There was, for example, in the nineteenth century the first female presidential candidate for the Presidency of the United States of America, who was called Victoria Woodhull. She did not do well in the election, being beaten by Ulysses Grant, who became the eighteenth president of the USA. The point about Victoria Woodhull was not her presidential aspirations; it was the publication of two books that she had written. The first of these was published in 1888 and titled *Stirpiculture, or the propagation of the human Race*. 'Stirpiculture' is a word that refers to the production of thoroughbred species or races. The second publication, produced in 1891, was *The Rapid Multiplication of the Unfit*, a title that speaks for itself. In this publication she suggested that if you want a better, superior people, then you should actively encourage the fit and healthy to breed and the unfit should be discouraged.

This philosophy crept into psychological texts of the time, where there were competitions to see which family could win as the fittest family, with as many members as possible. Unfortunately this was not always the happiest use of genetics. By 1930 about 24 US states had legislation that either required, or allowed, the sterilization of what was then described as the 'feeble minded'. A few books had turned into actual discriminating legislation; it is plainly evident how much care needs to be taken regarding these things.

One early form of social engineering took the form of castrating young male sopranos so that their voices did not drop at puberty. Indeed, the last known castrato died only in 1922. This process of castration was implicitly condoned by the church, most notably by the Catholic church. There were a number of very well known castrati and there were a number of famous castrati, for which composers of the day would write. These composers included Monteverdi, Handel and Mozart. But it was not entirely confined to that religion. The end of this barbaric practice happened with the election of the Pope, Leo XIII. He was originally called Vincenzo Gioacchino Pecci, became pope in 1878 and went on until his death in 1903, stopping the practice of castration for the vicarious pleasure of the listening public. Even though this practice was stopped for singers, it continued as a political practice in the name of eugenics.

A second form of this is the idea of a eunuch, a systematically brutalized individual. They were normally castrated as children with the aim of them becoming guards in harems of the Middle and Far East. They were also employed as chamberlains and were frequently very active in politics. This practice went on until well into the twentieth century. Humanity, with regards to eugenics, has always shown its darker side.

Eugenic societies held the belief that it was going to be possible to 'improve' humanity by the process of selective breeding of those individuals with what were thought to have 'good' qualities. What had gone unnoticed in this theory is something that a decent nineteenth-century gamekeeper would have been able to state – hybrid vigour.

It has become possible with selective breeding to produce a very wide range of animals and plants that are nothing like the progenitor species. We can look at this as either a process of augmented evolution, or messing with the natural forms of life. The DNA detectives are not in the position answering the question, but presenting us with the question by virtue of actions that take place in the laboratory. Yet we need to be more proactive in assessing whether what we do is right or wrong.

When DNA profiling was started, the Home Office of the UK explicitly stated that no profiling test would be carried out that could potentially diagnose a genetic disorder. This has not remained true. There is now ample scope for people to be discriminated against with regards to their genes – genes that are predisposed to illness, or obesity, or depression, or mental illness, to name only a few examples. Another potential problem is perhaps rather more serious. We all know of cases where one nationality takes a dislike to another and wages a war, of either nation against nation: genocide. Think how much worse it would be if it were possible to determine a racial origin from a DNA sample. Think about the information already tapped into a largely accessible database, or available with a network of extensive DNA identity cards. The 'unfit' are 'cleansed', remembering here that 'unfit' only means 'not the same'. This is something that has happened in the past, is happening now most probably, and will happen in the future.

There is one hugely interesting aspect to the idea of race in genetic terms – it appears not to exist. Given an individual you may think you could say whether they are African, European, Asian or any other broad category of humanity that you care to put a label on. A lot of the world has seen a blurring of racial boundaries, through migration and intermarriage and interbreeding. This is a slow and rather random process. If you were to take a complete range of humanity it might be hard to say where European features became Near Eastern, Middle Eastern and Far Eastern, and moving up from Africa the same situation arises. However, if you look at the genetic make-up of the people in these racial groups, there is no distinction to speak of.

Basically the idea that it is possible to determine race by genetics is not a viable proposition. One aspect that has been looked at is specific genetic markers, a marker being a section of DNA that can be clearly identified and analysed. This method was used to compare large groups of individuals from different parts of the world and gave different results depending on which particular markers were used. So sometimes there would appear to be two groups and sometimes four. The suspicion that was aroused by this is that for humans the concept of race does not exist: physical differences are only skin deep and in genetic terms we are all pretty much the same.

When it comes to the use of genetic data by governments within their own boundaries, this has not ceased and it is here that we can see best the importance of the idea of a benign state with regards to the advancing power of DNA and genetic science. It revolves around the idea that some genes are good and some genes are bad – two notions, both of which are flawed. If a gene was shown to be lethal in every case we could reasonably say that it was a bad gene. If, on the other hand, a gene was seen to be only lethal when an individual carried two copies of it we might be better off asking if there are any benefits to an individual carrying a single copy of a gene, being a carrier, to the population. It must always be remembered that genetics is not necessarily just about the individual; it is about populations and species. So why should recessive genes, lethal in a double dose, be tolerated, even perpetuated in a species? The simple answer is that for the most part we just do not know, but for some we have a good idea. So when a government decides, on eugenic grounds, what is and what is not good for its people we have to take seriously two things. The first is the human rights of the individuals concerned and the second is whether this action is going to be detrimental in the long term to a population, or humanity.

An example of this is the Chinese government who produced an edict in 1995, the broad meaning of which was that couples that had been diagnosed as having genetic disease should be forbidden to marry without either sterilization or long-term contraception. This is quite significant in the long term, for us all. So many genetic diseases are the result of spontaneous mutation. Consequently this government-imposed personal tragedy will not help any of us; just impoverish us all. But there is also another point about this: what if we eradicate all so-called disease genes? We simply do not know what genes are 'good' or 'bad'. An example of this is that group of genes which are generally called oncogenes. These are the

177

genes which, when switched on inappropriately, result in a cancer. So if we had tried to get rid of the oncogenes before we understood what their real use is we would have made a major mistake. What we refer to with the sinister name of oncogenes are in fact essential to development, both at the cellular, tissue, organ and organism level. No oncogenes, no life.

What the implications are for us all are wide reaching and not necessarily easy to appreciate. For example, if we were to remove a known mutation, without knowing why it is there in the first place, we could actually be storing up problems for ourselves. There are two examples of this that we know of.

The first of these is associated with cystic fibrosis, which is more common in the northern climates of Europe. This rather unpleasant disease is the result of the coming together of a carrier egg and carrier sperm, cystic fibrosis being a recessive condition. Carriers are unaffected; it is only the individuals who carry two copies of the defective gene that are affected. It is one of the commonest severe recessive disorders, appearing in approximately 1 in 2000 live births. Working backwards from this figure it is a straightforward calculation to work out that this meant a carrier frequency of about 5%, or 1 in 20, even before the gene was known and direct testing possible. So why should such a very debilitating condition be maintained at such a high frequency? For the investigation into the reason for this high rate we have to go back in time. It has been suggested that in the past, when tuberculosis was a common cause of death and disease, carriers of the mutation seem to be rather better protected against the disease than non carriers.

Perhaps a better example of protection of carriers against disease is found in the case of sickle-cell disease. A defective gene causes this condition; again it is recessive in that carriers are not generally affected, but people with two copies are severally affected. The red blood cells sickle as a change takes place in the red blood cells. This is not in itself what causes the trouble. The problems arise because the red blood cells that have collapsed to the sickle shape block up the small capillaries, which deprives the surrounding tissues of oxygen, which can eventually lead to organ failures and stroke.

The scale of the problem can be seen from the bare statistics. Each year, worldwide, there are approximately 100,000 children born each year with sickle-cell disease. In the UK there are between 5,000 and 6,000 patients, while in the USA there are 65,000 patients. The interesting thing about this is that virtually all of those affected are African in ethnic origin. Just as in cystic fibrosis, it is thought that carriers, but not affected, gain a degree of resistance to an infectious disease – in the case of sickle cell it is malaria.

These two examples indicate that even if we could remove all the defective genes we might well be causing more long-term problems for ourselves than we would solve. It also tells something else, that is, that when the astonishing power of DNA is brought to bear on a problem we have to tread very carefully or we might trip over our own cleverness and leave a legacy for future generations for which there is no answer. We may even create problems for individuals in the short term as well. For example, a modern, twenty-first century, conundrum has occurred over the cloning of individuals.

Those animals that have been cloned seem to have a shortened life expectancy, although it has to be said that this is not universally agreed on. It would be well to tread carefully with the idea of cloning a human being. At this stage we simply do not know enough about what goes on to risk a short and difficult life for an individual. It is not all negative, though. Processes devising new and better ways of treating individuals, even for infectious diseases, are moving forwards with genetic research that has already been looked at for its potential in forensic applications.

Genetic research involves single nucleotide polymorphisms (SNPs – pronounced snips). These have spawned the new science of pharmacogenetics. The importance of this new subject came from an observation that had been made in the 1950s. Just before administration of a general anaesthetic it is normal to give a pre-med relaxant. Unfortunately, when a pre-med was administered to some individuals they simply stopped breathing. These people had a mutation that did not allow them to clear the pharmaceutical from their system and this is what caused the problem.

Everyone knows someone, or it may even have happened to you personally, who has been given a prescription drug that has not 'agreed' with them. This is not uncommon.

There are about 3 billion prescriptions taken every year in the USA, of which there are as many as 2 million reported adverse reactions, and of these approximately 100,000 deaths, putting it as the fourth commonest cause of death. The only reasonable explanation of this difference in reaction is that the different individuals have different abilities to metabolize, or utilize, the same drug. So if it were possible to take a 'snap-shot' of an individual's capacity to deal with a specific drug a lot of discomfort and misery could be avoided. This is where SNP analysis comes in to the story. By relating a SNP to an adverse drug reaction it should be possible to stop prescribing drugs that are going to cause more problems for the patient than it will cure. By having a series of SNP probes arrayed on a chip, the equivalent of an electronic chip, a whole range of tests can be carried out in one go. What is more it is envisaged that this would not be done in some obscure laboratory but in the doctors consulting room, while you wait to find out which drug is best for you.

Of course, whenever genetic information is taken it is possible for it to be misused, especially when stored with individual names attached to it. We seem over the years to have attached more and more power to the state and less to individuals and in this sense state can also mean large and powerful businesses. It is often imagined that in times past an absolute monarch would have absolute power over his people, but today the power of the ruling classes has grown into new ways of control. If you wanted to identify a person back in the feudal days, it was necessary to have someone trusted who knew the individual, an identification parade of sorts. But now, with a massively complicated and interwoven infrastructure, it is possible for an anonymous individual working in a laboratory to identify the person from whom a sample originated, put a name to that person and later confirm the person arrested was the same person whose name they have and who left the crime scene sample. This is putting more power into the hands of the state because the whole process becomes progressively more impersonal and remote from democracy, from the people, from view. Government legislates and the bureaucracy of state implements is as impersonal a system as it is possible to get.

One of the more interesting conundrums about DNA and its uses is when large corporations, not states, get involved. Under these circumstances we need to tread very carefully and expect our state to protect us from the excesses, or potential excesses, of such companies. An example of this is the intention of the Icelandic government to allow a single licensed company to collect data on all those willing in the population, currently in excess of 270,000. The data would involve medical data and genetic data, but also genealogical information. Such a database would be an immense resource for tracking genes associated with disease, but with this level of information there was a great deal of concern expressed as to the number of safeguards so that the data could not be misused in the future. Such concerns prompted a large proportion of the scientists that would have been involved to turn their back on the project.

A similar project has been started in Umeå, the capital of northern Sweden. The entire region has a population of only 500,000. In comparison with the population density of Iceland, at three people for every square kilometre, Sweden has 19 people/sq km. The reason that this area of Sweden is being looked at is that there are considerable *founder effects* in homogenous populations around the capital. A founder effect describes a situation where an entire population starts from a very small number of individuals or families. For example, in the Azores almost the entire contemporary population can be traced back to one of 17 founding families. With the founder effect in northern Sweden there are a number of rare genetic diseases that can be found. Besides that, for the last 17 years the health care system has been offering free medical checks, including a blood test every ten years after the age of 30. Consequently there is a comprehensive register of both blood samples and lifestyle. Besides that information, parish records are accurate back to the fifteenth century, so it is possible to trace families with genetic diseases. It is the intention to compare the Swedish data with that gained from the Azores, where the genealogical data was also well documented, but in the Portuguese data it had been collected more by individuals and families, rather than as parish records. It is not uncommon to find a family bible of great antiquity in which there is space designed as a family register of births, marriages and deaths.

Another case of potential, or even in some respects current, misuse involves the insurance industry. This also requires an explanation of the way in which insurance works. One of the problems is that if you do have a test for a genetic disease there is some dispute as to whom should have access to the results. Insurance companies already tend to ask detailed questions regarding family history and whether tests for such things as HIV have been taken. With the increasing knowledge of genetic diseases, this also has ramifications for insurance companies, mainly because there are many conditions where having a disease gene does not mean that you will develop the condition. An example of this is one of the genes associated with breast cancer where 30% of those who have the gene remain perfectly healthy. This sort of data, if given to an insurance company, could result in grossly inflated premiums for perfectly healthy people, or even exclusion from health insurance entirely, so we could find a situation where those that need health insurance cannot get it, or afford it, and only those that are genetically healthy can get insurance.

This is a situation that can potentially be found in the USA, where the majority of health care is in the hands of private companies. Their insurance is based on a system called mutuality, where all those of similar risk are kept together and share the costs of their risks amongst them in their premiums. Consequently insurance becomes what is essentially a 'cherry picking' exercise with high-risk individuals not particularly welcome on a company's books.

The other usual method of insurance is called solidarity. In this system everybody shares the risk, and the premiums. This is the system generally found in the UK, but it is starting to move away from the idea with direct and known risks such as smoking and obesity. If insurance companies also started taking into account genetic information it would undermine the solidarity model at the expense of the individual. Since the time of the introduction of this equitable system of insurance the game of chance has been played out using actuarial tables, the number of people dying at any given age, it did not matter what of. This system works very well for all concerned. If the actuarial system becomes fractured it will only be to increase profits and result in the same situation as mutuality. Sadly in 1998 the UK government, against their own advisers, decided to allow genetic data to be used in assessing risks, even though I doubt they have the expertise to do so.

There is a worrying trend here: if insurance companies do not have adequate expertise to interpret genetic data, how much less expertise does the average person have in these matters? This is important because there it would seem that there is a development that bears on the question, the sale of do-it-yourself genetic tests. These can be purchased by post, over the internet or even in high street shops. These tests are designed to appeal to the worried, and as such could quite easily cause far more problems than could be easily imagined since some of the science is on very shaky ground. It is possible to find self-test kits that can suggest susceptibility to things like cancer, also single-gene conditions like cystic fibrosis.

One of the very real risks of doing a self-test is that an individual may become fatalistic and disturbed by a negative result, or unrealistically reassured by a positive result. This, of course, depends on the test being reliable – very few tests even when carried out in a professional laboratory will give a 100% accurate result. If an individual is, say, reassured that the risk of being susceptible to a tested condition, such as cancer, is zero, then an individual may think himself or herself immune to the problem and ignore symptoms until it is just too late. The alternative is also likely, a negative result may result in an individual effectively giving up on altering their lifestyle, which might be all that would have been required to avoid the problem all together, giving a long and healthy life.

Realistically any genetic test should be accompanied by proper counselling and only offered to those who are genuinely at risk. Even then it is important that any test is known to work and the problem here is that these tests are not regulated, as they are not medicines. One such is a test for the vitamin D receptor as an indicator of susceptibility to osteoporosis, the progressive weakening of bones due to lack of calcium. Vitamin D is essential for the incorporation of calcium in bones. This test originated in the USA but is sold by complementary therapists in the UK and there is a lot of contradictory evidence linking this gene to osteoporosis. Basically such a test is unlikely to tell you anything that can be used to help prevent osteoporosis. There is reluctance on the part of the manufacturers to state both the positive and negative published data about their tests. This is understandable in commercial terms but not entirely honest.

With many of these over-the-counter tests it is simply not enough to test a single gene and make a pronouncement based on the result. Many conditions are not controlled in a simple manner; inheritance is multi-factorial, which means that it is several genes and the environment that all influence the outcome. Perhaps even worse than simply testing for a gene some companies will then try and sell nutritional supplements to you as a method to stave off the suggested results of a spurious result.

Over-the-counter testing might, if not now, then in the future, have other ethical issues associated with it. This is the patenting of genes. If genetic testing becomes big business, then so does the patenting of the gene being tested, not just the test itself. Ostensibly this may seem reasonable, but grave doubts have been raised regarding patenting and licensing of genes and DNA sequences of any sort. The doubts revolve around the question as to whether patenting genes would impede research and the development of new medical treatment and devices.

Patenting of genes seems to fall into one of two categories. The first is made up of commercial companies, who tend to patent when ever possible to build up a wide range of potential products that have market value but also can be used to block competitors and give them a competitive edge. University and research institutions tend to look at the potential for commercial exploitation before filing a patent: they are not generally interested in restricting the use of the patent. These patents tend to be licensed, either to independent companies or spin-off companies specifically set up to make use of the discovery or invention. Sometimes a patent will be filed so that control is retained while commercial potential is investigated.

Whether the patenting of genes really does stifle research and innovation is difficult to be certain about, but it does make the reagents and equipment prohibitively expensive for geneticists and molecular biologists. That in itself will make it difficult to continue new research, but if the patent is of such potential commercial value it may not be possible to use the material at any price. This would most certainly put a stop to fundamental research, that may have resulted in a new and important development. The research put in by scientists over the world has led to books like this being written – where new technologies and tests have solved criminal and civil cases and made life better, or more efficient, or effected more just judgements, in real life.

That then has been an exploration of some of what the future might hold for the DNA detectives and their chosen field, DNA itself. It is indeed important to be very wary with a commodity and field as important as this. For it to fall into the wrong hands would be grievous indeed, yet as we have seen throughout this book, in the right hands it is a powerful tool. It has helped give forensic scientists and the law courts a new and exciting edge in the fight against crime. It has also, with recent developments, opened a new path for scientific endeavour. We will hear much more about DNA, but for the time being we've explored it in enough detail.

INDEX